CONTENTS

PREFACE

TO THE INSTRUCTOR

This study guide is designed to help your students master basic anatomy and physiology. It works in two ways.

First, the section of the preface, titled "To the Student," contains detailed instructions on:

How to achieve good grades in anatomy and physiology
How to read the textbook
How to use the exercises in this study guide
How to use visual memory as a learning tool
How to use mnemonic devices as learning aids
How to prepare for an examination
How to take an examination
How to find out why questions were missed on an examination

Second, the study guide itself contains features that facilitate learning. These features include the following:

1. **LEARNING OBJECTIVES**, designed to break down the process of learning into small units. The questions in this study guide have been developed to help the student master the learning objectives identified at the beginning of each chapter in the text. The guide is also sequenced to correspond to key areas of each chapter. A variety of questions have been prepared to cover the material effectively and expose the student to multiple learning approaches.

2. **CROSSWORD PUZZLES**, to encourage the use of new vocabulary words and emphasize the proper spelling of these terms.

3. **OPTIONAL APPLICATION QUESTIONS**, particularly targeted for the health occupations student, but appropriate for any student of anatomy and physiology because they are based entirely on information contained within the chapter.

4. **DIAGRAMS**, with key features marked by numbers for identification. Students can easily check their work by comparing the diagram in the workbook with the equivalent figure in the text.

5. **PAGE NUMBER REFERENCES**, in the answer sections. Each answer is keyed to the appropriate text page. Additionally, questions are grouped into specific topics that correspond to the text. Each major topic of the study guide provides references to specific areas of the text, so that students having difficulty with a particular grouping of questions have a specific reference area to assist them with remedial work. This is of great assistance to both instructor and student, because remedial work is made easier and more effective when the area of weakness is identified accurately.

These features should make mastery of the material a rewarding experience for both instructor and student.

Study Guide to Accompany

STRUCTURE & FUNCTION OF THE BODY

Ninth Edition

Prepared by
Linda Swisher, R.N., M.A.
Sarasota Vocational Technical Center
Sarasota, Florida

Mosby-Year Book, Inc.
St. Louis • Baltimore • Boston • Chicago • London • Philadelphia • Sidney • Toronto

Editor: Deborah Allen
Developmental Editor: Amy Winston
Production Editor: Donna Walls
Cover design: Gary Kaemmer

Previous edition copyrighted 1988

Printed in the United States of America

Mosby-Year Book, Inc.
11830 Westline Industrial Drive
St. Louis, MO 63146

ISBN 0-8016-64071

TO THE STUDENT

HOW TO ACHIEVE GOOD GRADES IN ANATOMY AND PHYSIOLOGY

This study guide is designed to help you help yourself to be successful in learning anatomy and physiology. Before you begin using the study guide, read the following suggestions. Students who understand effective study techniques and who have good study habits are successful students.

HOW TO READ THE TEXTBOOK

Keep up with the reading assignments. Read the textbook assignment prior to instructor covering the material in lecture. If you have failed to read the assignment beforehand, you will not grasp what the instructor is talking about in lecture. When you read, do the following:

1. As you finish reading a sentence, ask yourself if you understand it. If you do not, put a question mark in the margin by that sentence. If the instructor does not clear up the problem in lecture, ask him or her to explain it to you.

2. Do the learning objectives in the text. A learning objective is a specific task that you are expected to be able to do after you have read a chapter. It sets specific goals for the student and breaks down learning into small steps. It emphasizes the key points that the author is making in the chapter.

3. Underline and make notes in the margin to highlight key ideas, to mark something you need to reinforce at a later time, or to indicate things that you do not understand.

4. If you come to a word you do not understand, look it up in a dictionary. Write the word on one side of an index card and its definition on the other side. Carry these cards with you, and when you have a spare minute, use them in the way that you used flash cards to learn multiplication tables. If you do not know how to spell or pronounce a word, you will have a hard time remembering it.

5. Carefully study each diagram and illustration as you read. Many students ignore these aids. The author included them for a reason: to help students understand the material.

6. Summarize what you read. After finishing a paragraph, try to restate the main ideas. Do this again when you finish the chapter. Identify and review in your mind the main concepts of the chapter. Check to see if you are correct. In short, be an *active* reader. Do not just stare at a page or read it superficially.

Finally, attack each unit of learning with a positive mental attitude. Motivation and perseverance are prime factors in achieving successful grades. The combination of your instructor, the text, the study guide, and your dedicated work will lead to success for you in anatomy and physiology.

HOW TO USE THE EXERCISES IN THIS STUDY GUIDE

After you have read a chapter and learned all the new words, begin working with the study guide. Read the overview of the chapter, which summarizes the main points.

Familiarize yourself with the Topics for Review section, which repeats the learning objectives outlined in the text. Complete the questions and diagrams in the study guide. The questions are sequenced to follow the chapter outline and headings and are divided into small sections to facilitate learning. A variety of questions is offered throughout the study guide to help you cover the material effectively. The following examples are among the exercises that have been included to assist you.

Multiple Choice Questions

Multiple choice questions will have only one correct answer out of several possibilities for you to select. There are two types of multiple choice questions that you may not be acquainted with:

1. "None of the above is correct" questions. These questions test your ability to recall rather than recognize the correct answer. You would select the "none of the above" choice only if all the other choices in that particular question were incorrect.

2. Sequence questions. These questions test your ability to arrange a list of structures in the correct order. In this type of question you are asked to determine the sequence of the structures given in the various choices, and then you are to select the structure listed that would be third in that sequence, as in this example.

   Which one of the following structures would be the third through which food would pass?

   a. Stomach
   b. Mouth
   c. Large intestine
   d. Esophagus
   e. Anus

   The correct answer would be a.

Matching Questions

Matching questions ask the student to select the correct answer from a list of terms and to write the answer in the space provided.

True or False

True or false questions ask you to write T in the answer space, if you agree with that statement. If you disagree with the statement, you will circle the incorrect word(s) and write the correct word(s) in the answer space.

Identify the Incorrect Term

In questions that ask you to identify the incorrect term, three words are given that relate to each other in structure or function, and one more word is included that has no relationship, or has an opposing relationship to the other three terms. You are to circle the term that does not relate to the other terms. An example might be: iris, cornea, stapes, retina. You would circle the word *stapes* because all other terms refer to the eye.

Fill in the Blanks

Fill-in-the-blank questions ask you to recall missing word(s) and insert it or them into the answer blank(s). These questions may be sentences or paragraphs.

Application

Application questions ask you to make judgments about a situation based on the information in the chapter. These questions may ask you how you would respond to a situation or to suggest a possible diagnosis for a set of symptoms.

Charts

Several charts have been included that correspond to figures in the text. Areas have been omitted so that you can fill them in and test your recall of these important areas.

Identification

The study guide includes word find puzzles that allow you to identify key terms in the chapter in an interesting and challenging way.

Crossword Puzzles

Vocabulary words from the New Words section at the end of each chapter of the text have been developed into crossword puzzles. This not only encourages recall, but also proper spelling. Occasionally, an exercise uses scrambled words to encourage recall and spelling.

Labeling Exercises

Labeling exercises present diagrams with parts that are not identified. For each of these diagrams you are to print the name of each numbered part on the appropriately numbered line. You may choose to further distinguish the structures by coloring them with a variety of colors. After you have written down the names of all the structures to be identified, check your answers. When it comes time to review before an examination you can place a sheet of paper over the answers that you have already written on the lines. This procedure will allow you to test yourself without peeking at the answers.

After completing the exercises in the study guide, check your answers. If they are not correct, refer to the page listed with the answer, and review it for further clarification. If you still do not understand the question or the answer, ask your instructor for further explanation.

If you have difficulty with several questions from one section, refer to the pages given at the end of the section. After reviewing the section, try to answer the questions again. If you are still having difficulty, talk to your instructor.

HOW TO USE VISUAL MEMORY

Visual memory is another important tool in learning. If I asked you to picture an elephant in your mind, with all its external parts labeled, you could do that easily. Visual memory is a powerful key to learning. Whenever possible, try to build a memory picture. Remember, a picture is worth a thousand words.

Visual memory works especially well with the sequencing of items, such as circulatory pathways and the passage of air or food. Students who try to learn sequencing by memorizing a list of words do poorly on examinations. If they forget one word in the sequence, then they will forget all the words after the forgotten one as well. However, with a memory picture you can pick out the important features.

HOW TO USE MNEMONIC DEVICES

Mnemonic devices are little jingles you memorize to help you remember things. If you make up your own, they will stick with you longer. Here are three examples of such devices:

"On Old Olympus' towering tops a Finn and German viewed some hops." This one is used to remember the cranial nerves; each word begins with the same letter as does the name of one of the nerves.

"C. Hopkins CaFe where they serve Mg NaCl." This mnemonic reminds you of the chemical symbols for the biologically important electrolytes.

"Roy G. Biv." This mnemonic helps you remember the colors of the visible light spectrum.

HOW TO PREPARE FOR AN EXAMINATION

Prepare far in advance for an examination. Actually, your preparation for an examination should begin on the first day of class. Keeping up with your assignments daily makes the final preparation for an examination much easier. You should begin your final preparation at least three nights before the test. Last-minute studying usually means poor results and limited retention.

1. Make sure that you understand and can answer all of the learning objectives for the chapter on which you are being tested.

2. Review the appropriate questions in this study guide. Reviewing is something that you should do after every class and at the end of every study session. It is important to keep going over the material until you have a thorough understanding of the chapter and rapid recall of its contents. If review becomes a daily habit, studying for the actual examination will not be difficult. Go through each question and write down an answer. Do the same with the labeling of each structure on the appropriate diagrams. If you have already done this as part of your daily review, cover the answers with a piece of paper and quiz yourself again.

3. Check the answers that you have written down against the correct answers in the back of the study guide. Go back and study the areas in the text that refer to questions that you missed and then try to answer those questions again. If you still cannot answer a question or label a structure correctly, ask your instructor for help.

4. Ask yourself as you read a chapter what questions you would ask if you were writing a test on that unit. You will most likely ask yourself many of the questions that will show up on your examinations.

5. Get a good night's sleep before the test. Staying up late and upsetting your biorhythms will only make you less efficient during the test.

HOW TO TAKE AN EXAMINATION

The Day of the Test

1. Get up early enough to avoid rushing. Eat appropriately; your body needs fuel, but a heavy meal just before a test is not a good idea.

2. Keep calm. Briefly look over your notes. If you have prepared for the test properly, there will be no need for last-minute cramming.

3. Make certain that you have everything you need for the test — pens, pencils, test sheets, and so forth.

4. Allow enough time to get to the examination site. Missing your bus, getting stuck in traffic, or being unable to find a parking space will not put you in a good frame of mind to do well on the examination.

During the Examination

1. Pay careful attention to the instructions for the test.

2. Note any corrections.

3. Budget your time so that you will be able to finish the test.

4. Ask the instructor for clarification if you do not understand a question or an instruction.

5. Concentrate on your own test paper and do not allow yourself to be distracted by others in the room.

Hints for Taking a Multiple Choice Test

1. Read each question carefully. Pay attention to each word.

2. Cross out obviously wrong choices and think about those that are left.

3. Go through the test once, quickly answering the questions you are sure of. Then go back over the test and answer the rest of the questions.

4. Fill in the answer spaces completely and make your marks heavy. Erase completely if you make a mistake.

5. If you must guess, stick with your first hunch. Most often, students will change right answers to wrong ones.

6. If you will not be penalized for guessing, do not leave any blanks.

Hints for Taking an Essay Test

1. Budget time for each question.

2. Write legibly and try to spell words correctly.

3. Be concise, complete, and specific. Do not be repetitious or long-winded.

4. Organize your answer in an outline which helps not only the student but also the person who grades the test.

5. Answer each question as thoroughly as you can, but leave some room for possible additions.

Hints for Taking a Laboratory Practical Examination

Students have a hard time with this kind of test. Visual memory is very important here. To put it simply, you must be able to identify every structure you have studied. If you are unable to identify a structure, then you will be unable to answer any questions about that structure.

Possible questions that could appear on an examination of this type might include:

1. Identification of a structure, organ, feature.

2. Function of a structure, organ, feature.

3. Sequence questions for air flow, passage of food, urine, and so forth.

4. Disease questions (e.g., if an organ fails, what disease will result?).

HOW TO FIND OUT WHY QUESTIONS WERE MISSED ON AN EXAMINATION

After the Examination

Go over your test after it has been scored to see what you missed and why you missed it. You can pick up important clues that will help you on future examinations. Ask yourself these questions:

1. Did I miss questions because I did not read them carefully?

2. Did I miss questions because I had gaps in my knowledge?

3. Did I miss questions because I could not determine scientific words?

4. Did I miss questions because I did not have good visual memory of things?

Be sure to go back and learn the things you did not know. Chances are these topics will come up on the final examination.

Your grades in other classes as well will improve greatly when you apply these study methods. Learning should be fun. With these helpful hints and this study guide you should be able to achieve the grades you desire. Good luck!

ACKNOWLEDGEMENTS

I wish to express my appreciation to the staff of Mosby-Year Book, Inc., especially Deborah Allen, Laura Edwards and Amy Winston for opening this door. My continued admiration and thanks to Gary Thibodeau for another outstanding edition of his text. My gratitude to the reviewers, Eugene R. Volz, M.A., Ann Scott, M.A., Anna M. Strand, B.S.N. and Denise L. Kampfhenkel, B.S.N. Your time and dedication to science education will hopefully create a better quality of health care for the future.

Special thanks to Alice Tatakis for her perseverance and enthusiasm while transposing the written word to typed script.

A. Christine Payne and her computer combined efforts to produce crossword puzzles and word finds for the units. Her creativity added the variety necessary to stimulate the learning process.

To my mother, Randy, Di, Deb, Dianne, the Golden Girls and Bill my thanks for your assistance and constant support throughout this project and my life.

Finally, to my daughter, Amanda, this book is dedicated. You and I know the many reasons.

Linda Swisher, RN, M.A.

CHAPTER 1

An Introduction to the Structure and Function of the Body

A command of terminology is necessary for a student to be successful in any area of science. This chapter defines the terminology and concepts that are basic to the field of anatomy and physiology. Building a firm foundation in these language skills will assist you with all future chapters.

The study of anatomy and physiology involves the structure and function of an organism and the relationship of its parts. It begins with a basic organization of the body into different structural levels. Beginning with the smallest level (the cell) and progressing to the largest, most complex level (the system), this chapter familiarizes you with the terminology and the levels of organization needed to facilitate the study of the body as a part or as a whole.

It is also important to be able to identify and to describe specific body areas or regions as we progress in this field. The anatomical position is used as a reference position when dissecting the body into planes, regions, or cavities. The terminology defined in this chapter allows you to describe the areas efficiently and accurately.

Finally, the process of homeostasis is reviewed. This state of relative constancy in the chemical composition of body fluids is necessary for good health. In fact, the very survival of the body depends on the successful maintenance of homeostasis.

TOPICS FOR REVIEW

Before progressing to Chapter 2, you should have an understanding of the structural levels of organization; the planes, regions, and cavities of the body; the terminology used to describe these areas, and the concept of homeostasis as it relates to the survival of the species.

STRUCTURAL LEVELS OF ORGANIZATION

Match the term on the left with the proper selection on the right.

Group A

____ 1. Organism
____ 2. Cells
____ 3. Tissue
____ 4. Organ
____ 5. Systems

a. Many similar cells that act together to perform a common function
b. The most complex units that make up the body
c. A group of several different kinds of tissues arranged to perform a special function
d. Denotes a living thing
e. The smallest "living" units of structure and function in the body

Group B

____ 6. Dorsal
____ 7. Ventral
____ 8. Thoracic
____ 9. Pelvic
____ 10. Diaphragm

a. Chest cavity
b. Area below the hipbone
c. Toward the back
d. Separates the thoracic cavity from the abdominal cavity
e. Toward the belly

▸ If you have had difficulty with this section, review pages 1-6. ◂

SOME WORDS USED IN DESCRIBING BODY STRUCTURES, PLANES, OR SECTIONS

Fill in the crossword puzzle.

11. Upper or above
12. Lower or below
13. Horizontal plane
14. Front (abdominal side)
15. Toward the side of the body
16. Toward the midline of the body
17. Farthest from the point of origin of a body point

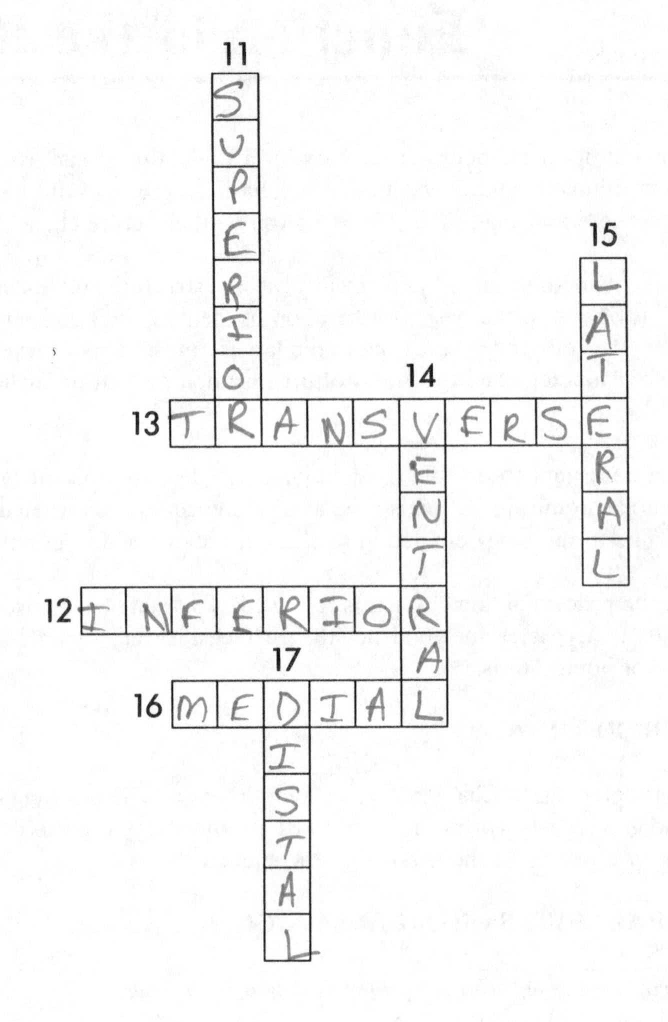

Circle the correct answer.

18. The bladder is (superior or inferior) to the transverse colon.

19. The nose is located on the (anterior or posterior) surface of the body.

20. The lungs lie (medial or lateral) to the heart.

21. The elbow lies (proximal or distal) to the forearm.

22. The skin is (superficial or deep) to the muscles below it.

23. A midsagittal plane divides the body into (equal or unequal) parts.

24. A frontal plane divides the body into (anterior and posterior or superior and inferior) sections.

25. A transverse plane divides the body into (right and left or upper and lower) sections.

26. The liver lies in the (right upper quadrant or right lower quadrant).

27. In the anatomical position the palms of the hands are pointing (forward or backward).

▸ If you have had difficulty with this section, review pages 5-9. ◂

BODY REGIONS

Circle the one that does not belong.

28. Axial	Head	Trunk	Extremities
29. Axillary	Cephalic	Brachial	Antecubital
30. Frontal	Orbital	Plantar	Nasal
31. Erect	Anatomical position	Feet forward	Eyes closed
32. Inguinal	Navel	Plantar	Mammary
33. Carpal	Femoral	Plantar	Pedal
34. Thoracic	Dorsal	Gluteal	Popliteal

▸ If you have had difficulty with this section, review pages 9-11. ◂

SOME BASIC FACTS ABOUT BODY FUNCTIONS

Fill in the blanks.

35. ______________ depends on the body's ability to maintain or restore homeostasis.

36. "Homeostasis" is the term used to describe the relative constancy of the body's ______________.

37. In the absence of homeostasis, body CO2 levels would ______________.

38. Changes and functions that occur during the early years are called ______________

39. Changes and functions that occur after young adulthood are called ______________.

40. Homeostatic control mechanisms are categorized as either ______________ or ______________ feedback loops.

41. Negative feedback loops are ______________ mechanisms.

42. Positive feedback control loops are ______________.

▸ If you have had difficulty with this section, review pages 12-13. ◂

APPLYING WHAT YOU KNOW

43. Mrs. Fagan has had an appendectomy. The nurse is preparing to change the dressing. She knows that the appendix is located in the right iliac inguinal region, the distal portion extending at an angle into the hypogastric region. Place an X on the diagram where the nurse will place the dressing.

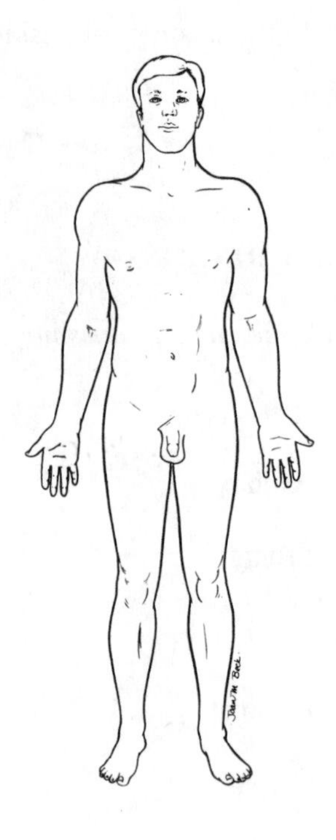

44. Mrs. Suchman noticed a lump in her breast. Dr. Reeder noted on her chart that a small mass was located in the left breast medial to the nipple. Place an X where Mrs. Suchman's lump would be located.

45. Jeff was injured in a bicycle accident. X-ray films revealed that he had a fracture of the right patella. A cast was applied beginning at the distal femoral region and extending to the pedal region. Place an X where Jeff's cast begins and ends.

DID YOU KNOW?

Many animals produce tears but only humans weep as a result of emotional stress.

DORSAL AND VENTRAL BODY CAVITIES

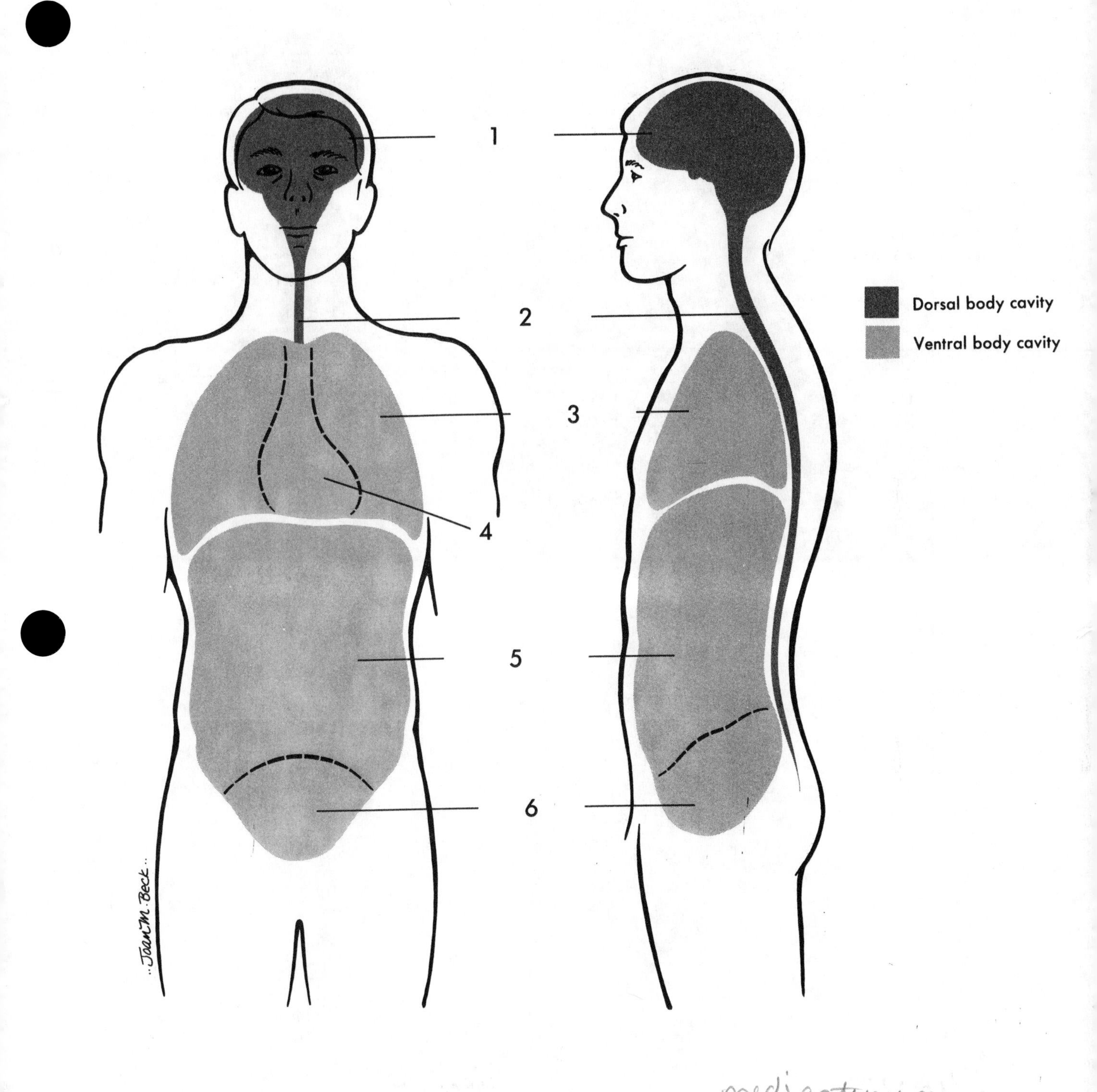

1. cranial cavity
2. spinal cavity
3. thoracic cavity
4. pleural cavity (mediastinum)
5. abdominal
6. pelvic

DIRECTIONS AND PLANES OF THE BODY

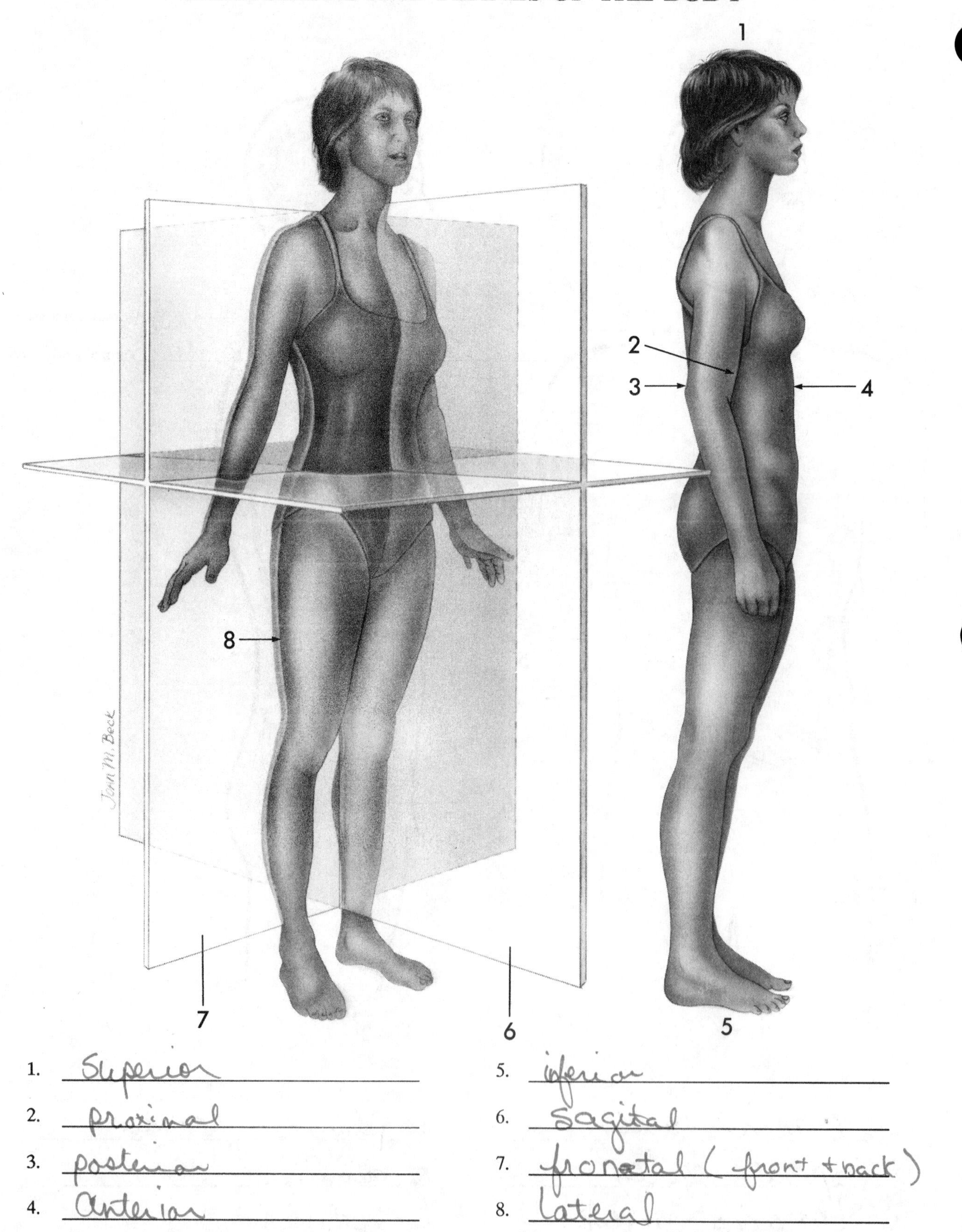

1. Superior
2. proximal
3. posterior
4. Anterior
5. inferior
6. Sagital
7. fronatal (front + back)
8. lateral

REGIONS OF THE ABDOMEN

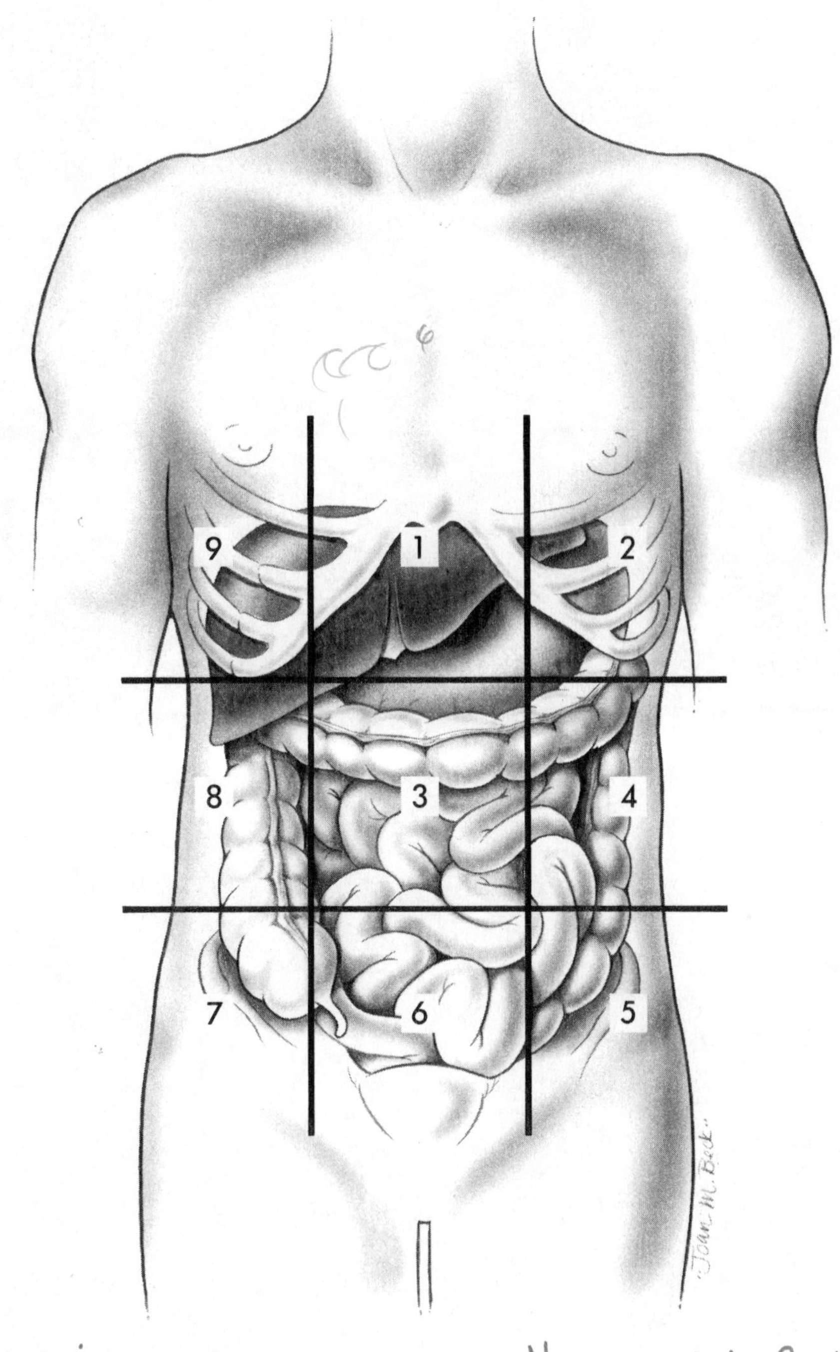

1. Epigastric
2. Left hypochondriac region
3. Umbilical region
4. Left Lumbar region
5. Left iliac (inguinal) region
6. Hypogastric Region
7.
8.
9.

CHAPTER 2

Cells and Tissues

Cells are the smallest structural units of living things. Therefore, because we are living, we are made up of a mass of cells. Human cells can be viewed only microscopically; they vary in shape and size. The three main parts of a cell are the cytoplasmic membrane, the cytoplasm, and the nucleus. As you review the chapter on cells, you will be amazed at the cells' resemblance to the body as a whole. You will identify miniature circulatory systems, reproductive systems, digestive systems, power plants much like the muscular system, and many others that will aid in your understanding of these systems in future chapters.

Cells, just like humans, depend on water, food, gases, elimination of wastes, and numerous other substances. The movement of these substances in and out of the cells is accomplished by two primary methods: passive transport processes and active transport processes. In passive transport processes no cellular energy is required to effect movement through the cell membrane. However, in active transport cellular energy is required to effect movement through the cell membrane.

Cell reproduction completes the study of cells. A basic explanation of DNA, "the hereditary molecule," gives us a proper respect for the capability of the cell to transmit physical and mental traits from generation to generation. Reproduction of the cell, mitosis, is a complex process requiring several stages. These stages are outlined and diagramed in the text to facilitate learning.

This chapter concludes with the discussion of tissues. The four main types of tissues, epithelial, connective, muscle, and nervous, are reviewed. The characteristics, location, and function of these tissues are necessary to complete your understanding of the next structural level of organization.

TOPICS FOR REVIEW

Before progressing to Chapter 3, you should have an understanding of the structure and function of the smallest living unit in the body—the cell. Your review should also include the methods by which substances are moved through the cell membrane and the stages necessary for the cell to reproduce. The study of this chapter is completed with an understanding of tonicity and body tissues and the function they perform in the body.

CELLS

Match the term on the left with the proper selection on the right.

Group A

__c__ 1. Cytoplasm	a. Component of plasma membrane
__e__ 2. Plasma membrane	b. Controls reproduction of the cell
__a__ 3. Cholesterol	c. "Living matter"
__b__ 4. Nucleus	d. Paired organelles
__d__ 5. Centrioles	e. Surrounds cells

Cell Structure

Group B

____ 6. Ribosomes
____ 7. Endoplasmic reticulum
____ 8. Mitochondria
____ 9. Lysosomes
____ 10. Golgi apparatus

a. "Power plants"
b. "Digestive bags"
c. "Carbohydrate producing and packaging factory"
d. "Protein factories"
e. Miniature "circulatory system"

Fill in the blanks.

11. A system where measurement of length is based on the meter is known as the ______________________.

12. A procedure performed prior to transplanting an organ from one individual to another is ______________________.

13. Fine, hairlike extensions found on the exposed or free surfaces of some cells are called ______________________.

14. This organelle is distinguished by the fact that it has two types. It may be either smooth or rough ______________________.

15. ______________________ are usually attached to rough endoplasmic reticulum and produce enzymes and other protein compounds.

16. The ______________________ provide energy-releasing chemical reactions that go on continuously.

17. The organelles that can digest and destroy microbes that invade the cell are called ______________________.

18. Mucus is an example of a product manufactured by the ______________.

19. These rod-shaped structures, ______________________, play an important role during cell division.

20. ______________________ are threadlike structures made up of proteins and DNA.

▸ If you have had difficulty with this section, review pages 18-23. ◂

MOVEMENT OF SUBSTANCES THROUGH CELL MEMBRANES

Circle the correct choice.

21. The energy required for active transport processes is obtained from:

a. ATP
b. DNA
c. Diffusion
d. Osmosis

22. An example of a passive transport process is:

a. Permease system
b. Phagocytosis
c. Pinocytosis
d. Diffusion

23. Movement of substances from a region of high concentration to a region of low concentration is known as:

a. Active transport
b. Passive transport
c. Cellular energy
d. Concentration gradient

24. Osmosis is the ______ of water across a selectively permeable membrane.

a. Filtration
b. Equilibrium
c. Active transport
d. Diffusion

25. ______ involves the movement of solutes across a selectively permeable membrane by the process of diffusion.

a. Osmosis
b. Filtration
c. Dialysis
d. Phagocytosis

26. A specialized example of diffusion is:

a. Osmosis
b. Permease system
c. Filtration
d. All of the above

27. This movement always occurs down a hydrostatic pressure gradient.

a. Osmosis
b. Filtration
c. Dialysis
d. Facilitated diffusion

28. The uphill movement of a substance through a living cell membrane is:

a. Osmosis
b. Diffusion
c. Active transport process
d. Passive transport process

29. The sodium pump is an example of this type of movement.

a. Osmosis
b. Pinocytosis
c. Permease system
d. Diffusion

30. An example of a cell that uses phagocytosis is the:

a. White blood cell
b. Red blood cell
c. Muscle cell
d. Bone cell

31. A saline solution that contains a higher concentration of salt than living red blood cells would be:

a. Hypotonic
b. Hypertonic
c. Isotonic
d. Homeostatic

32. A red blood cell becomes engorged with water and will eventually lyse, releasing hemoglobin into the solution. This solution is ______ to the red blood cell.

a. Hypotonic
b. Hypertonic
c. Isotonic
d. Homeostatic

▸ If you have had difficulty with this section, review pages 24-26. ◂

CELL REPRODUCTION

Circle the one that does not belong.

33. DNA	Adenine	Uracil	Thymine
34. Complementary base pairing	Guanine	RNA	Cytosine
35. Anaphase	Specific sequence	Gene	Base pairs
36. RNA	Ribose	Thymine	Uracil
37. Double helix	Mitosis	DNA	Replication
38. Cleavage furror	Anaphase	Prophase	2 daughter cells
39. "Resting"	Prophase	Interphase	DNA replication
40. Identical	2 nuclei	Telophase	Metaphase
41. Metaphase	Prophase	Telophase	Gene

▸ If you have had difficulty with this section, review pages 27-30. ◂

TISSUES

42. *Fill in the missing area.*

TISSUE	LOCATION	FUNCTION
Epithelial		
1. Simple squamous	1a. Alveoli of lungs	1a.
	1b. Lining of blood and lymphatic vessels	1b.
2. Stratified squamous	2a.	2a. Protection
	2b.	2b. Protection
3. Simple columnar	3.	3. Protection, secretion, absorption
4.	4. Urinary bladder	4. Protection
5. Pseudostratified	5.	5. Protection

Connective		
1. Areolar	1.	1. Connection
2.	2. Under skin	2. Protection; insulation
3. Dense fibrous	3. Tendons; ligaments; fascia, scar tissue	3.
4. Bone	4.	4. Support, protection
5. Cartilage	5.	5. Firm but flexible support
6. Blood	6. Blood vessels	6.
7.	7. Red bone marrow	7. Blood cell formation
Muscle		
1. Skeletal (striated voluntary)	1.	1. Movement of bones
2.	2. Wall of heart	2. Contraction of heart
3. Smooth	3.	3. Movement of substances along ducts; change in diameter of pupils and shape of lens; "gooseflesh"
Nervous		
1.	1.	1. Irritability, conduction

▸ If you have had difficulty with this section, review Table 2-3 and page 31. ◂

APPLYING WHAT YOU KNOW

43. Mr. Fee's boat had capsized, and he was stranded on a deserted shoreline for 2 days without food or water. When found, he had swallowed a great deal of seawater. He was taken to the emergency room in a state of dehydration. In the space to the right, draw the appearance of the red blood cells as they would appear to the laboratory technician.

44. The nurse was instructed to dissolve a pill in a small amount of liquid medication. As she dropped the capsule into the liquid, she was interrupted by the telephone. On her return to the medication cart, she found the medication completely dissolved and apparently scattered evenly throughout the liquid. This phenomenon did not surprise her since she was aware from her knowledge of cell transport that ____________________ had created this distribution.

45. Mrs. Henion has emphysema and has been admitted to the hospital unit with oxygen per nasal cannula. Emphysema destroys the tiny air sacs in the lungs, reducing the diffusion of oxygen into the blood. These tiny air sacs, alveoli, are formed by what type of tissue?

46. Merrily was 5'4" and weighed 115 lbs. She appeared very healthy and fit, yet her doctor advised her that she was "overfat." What might be the explanation for this assessment?

DID YOU KNOW?

The largest single cell in the human body is the female sex cell, the ovum. The smallest single cell in the human body is the male sex cell, the sperm.

CELLS AND TISSUES

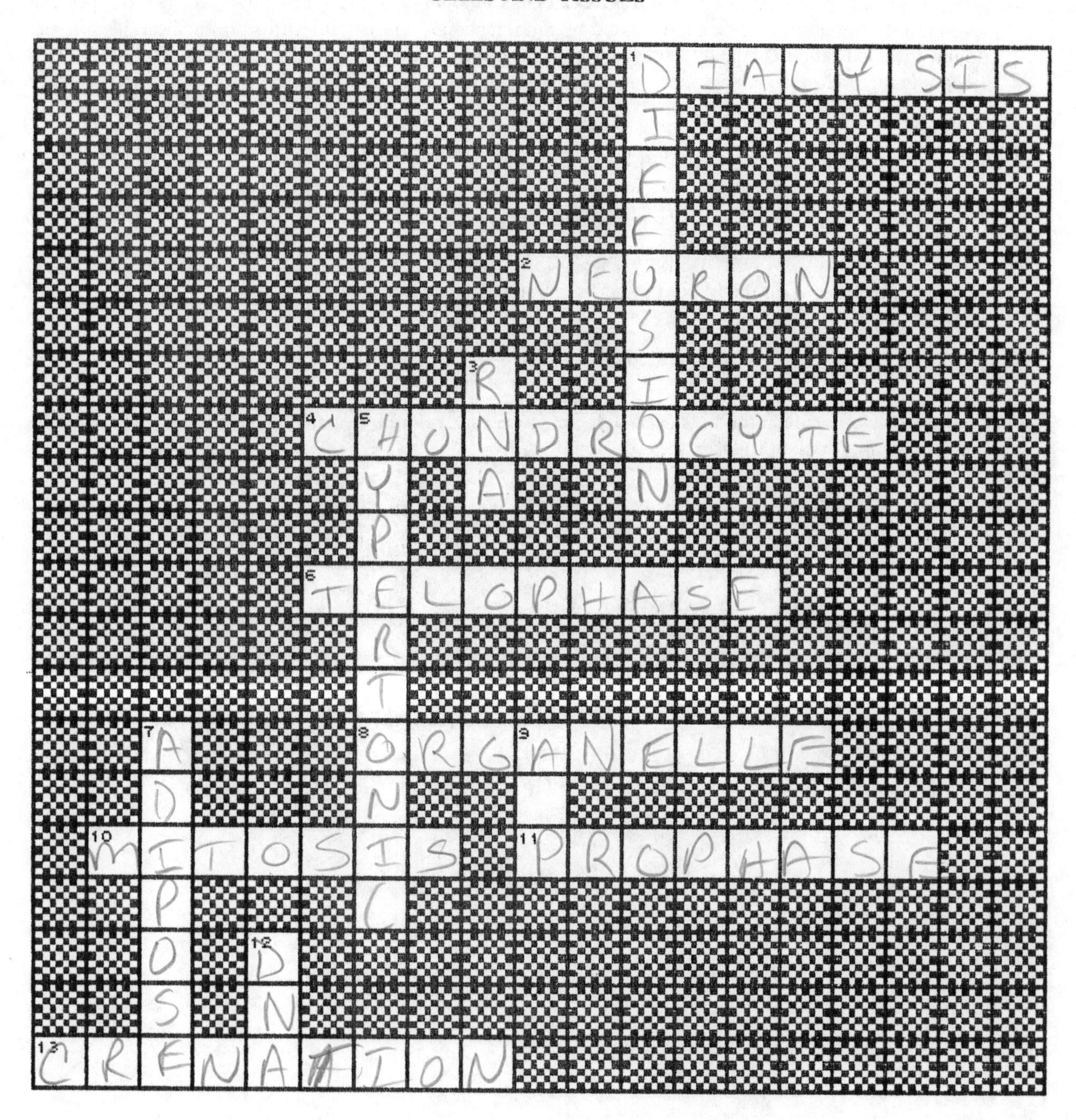

Across

1. Specialized example of diffusion
2. Nerve cell
4. Cartilage cell
6. Last stage of mitosis
8. Cell organ
10. Reproduction process of most cells
11. First stage of mitosis
13. Shriveling of cell due to water withdrawal

Down

1. Occurs when substances scatter themselves evenly throughout an available space
3. Ribonucleic acid (abbreviation)
5. Having an osmotic pressure greater than that of the solution of which it is compared
7. Fat
9. Energy source for active transport
12. Chemical "blueprint" of the body (abbreviation)

CELL STRUCTURE

21

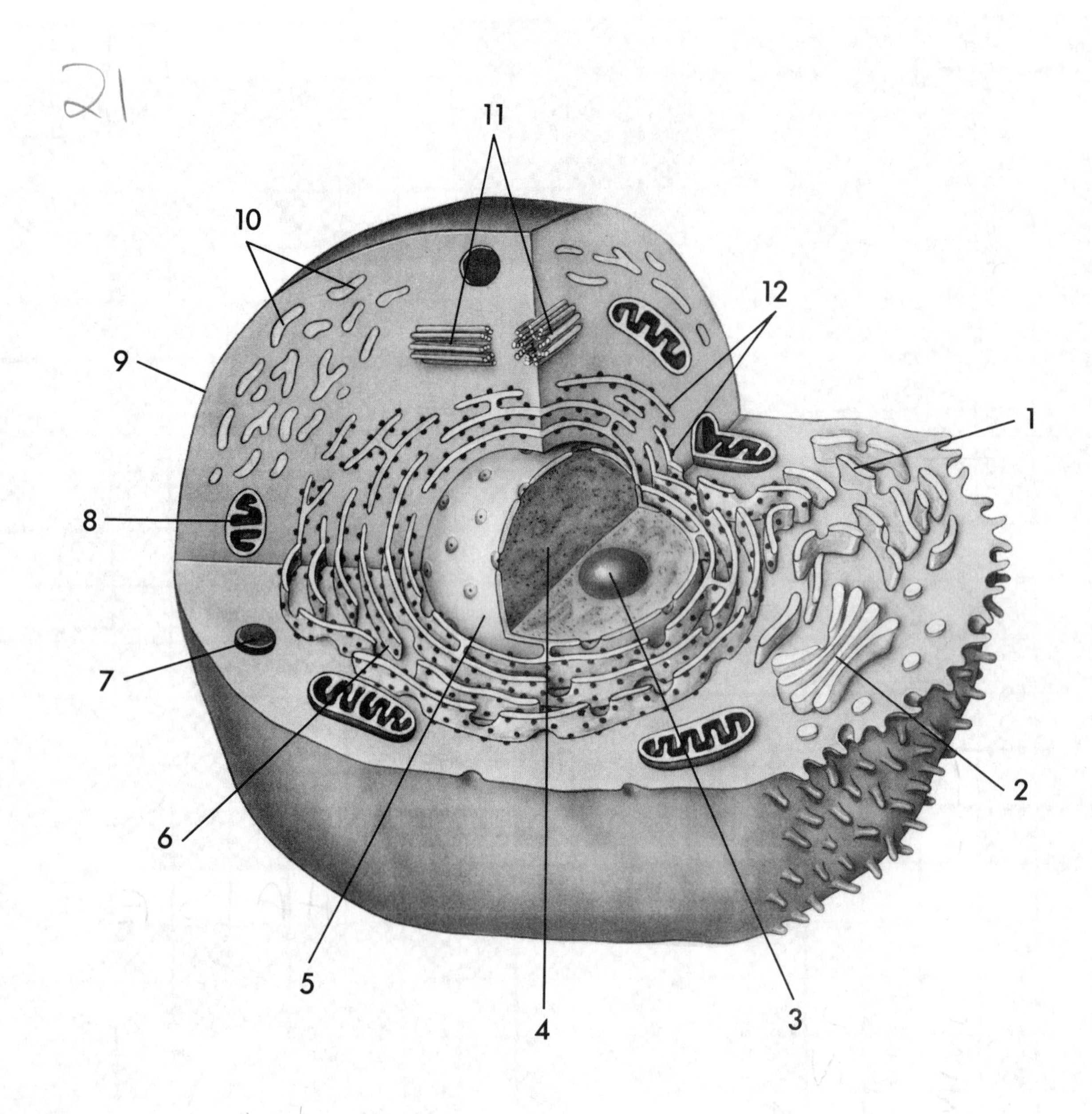

1. Smo
2. Golgi Apparatus
3. Nucleolus
4.
5.
6.
7.
8.
9.
10.
11. Centriole S
12.

MITOSIS

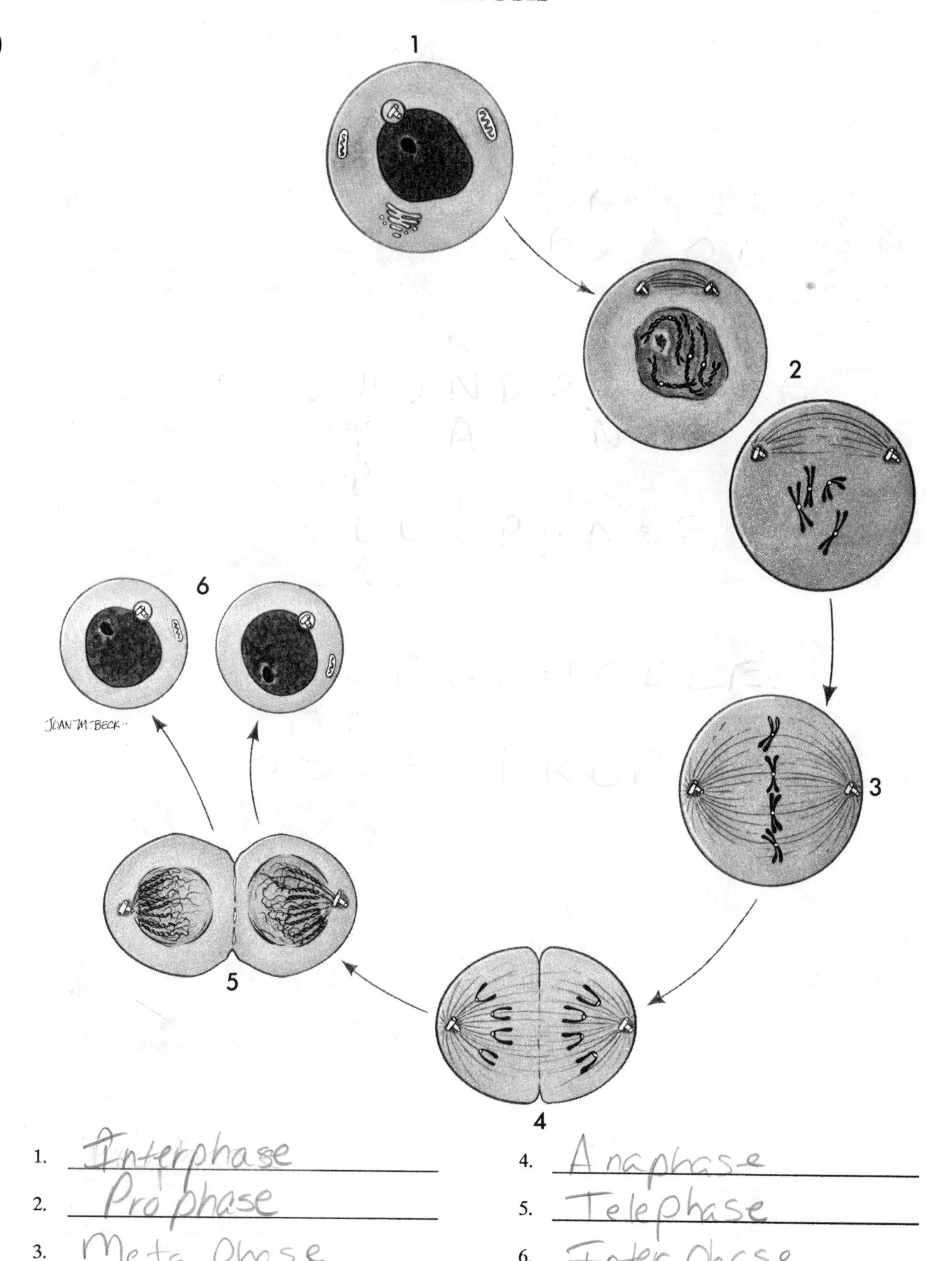

1
2
3
4
5
6
JOAN M. BECK
1. Interphase
2. Prophase
3. Meta phase
4. Anaphase
5. Telephase
6. Interphase

TISSUES

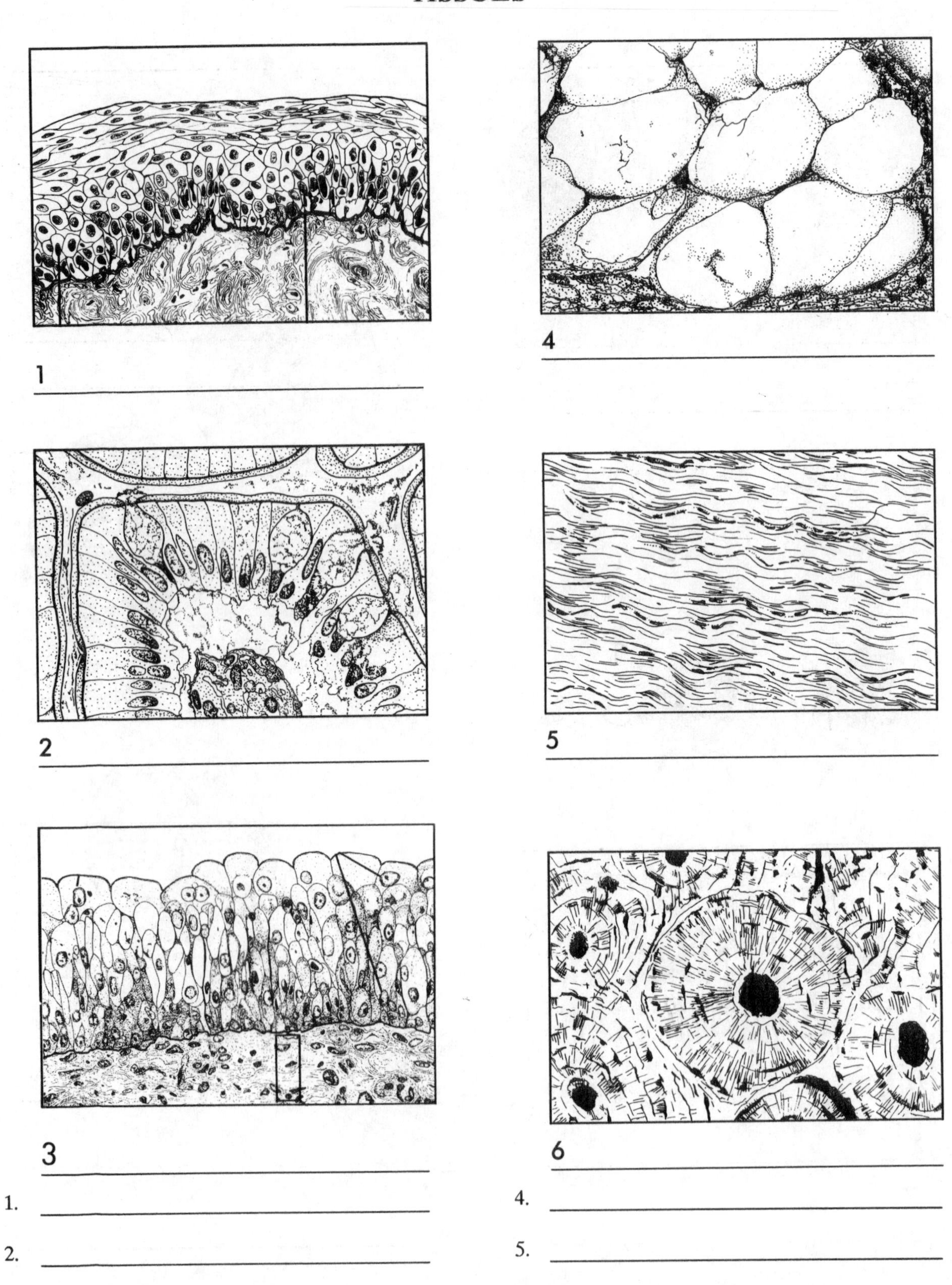

1. ______________________

2. ______________________

3. ______________________

4. ______________________

5. ______________________

6. ______________________

TISSUES

(Continued)

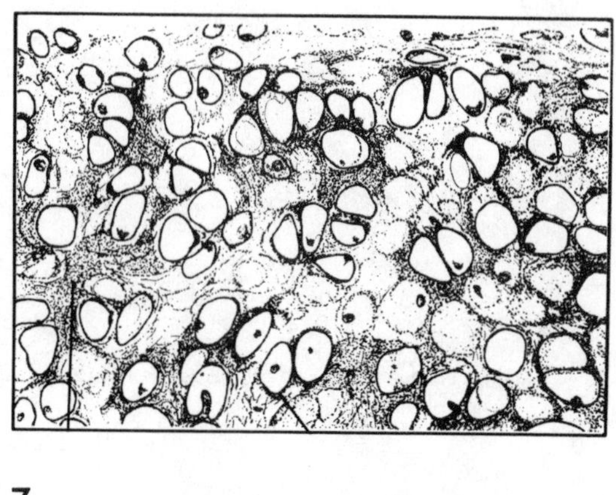

7 ______________________

10 ______________________

8 ______________________

11 ______________________

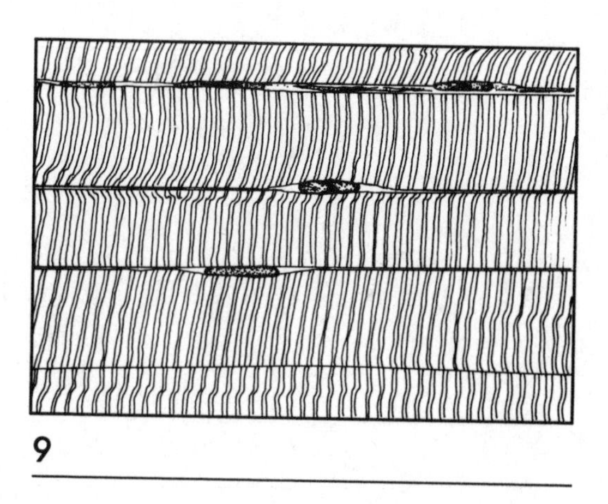

9 ______________________

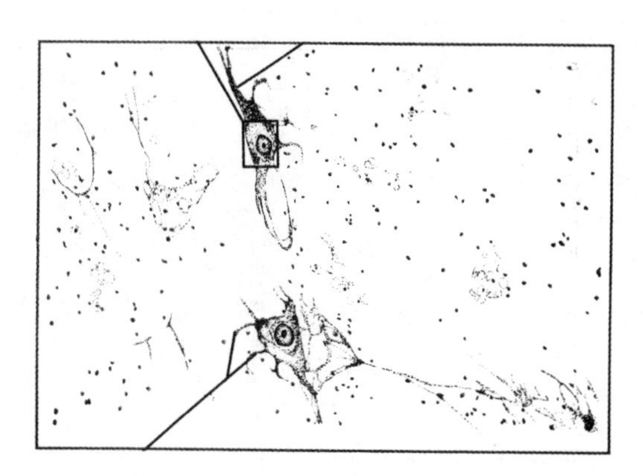

12 ______________________

7. ______________________

8. ______________________

9. ______________________

10. ______________________

11. ______________________

12. ______________________

CHAPTER 3

Organ Systems of the Body

A smooth-running automobile is the result of many systems working harmoniously. The engine, the fuel system, the exhaust system, the brake system, and the cooling system are but a few of the many complex structural units that the automobile as a whole relies on to keep it functioning smoothly. So it is with the human body. We, too, depend on the successful performance of many individual systems working together to create a healthy human being.

When you have completed your review of the 11 major organ systems and the organs that make up these systems, you will find your understanding of the performance of the body as a whole much more meaningful.

TOPICS FOR REVIEW

Before progressing to Chapter 4, you should have an understanding of the 11 major organ systems and be able to identify the organs that are included in each system.

ORGAN SYSTEMS OF THE BODY

Match the term on the left with the proper selection on the right.

Group A

___ 1.	Integumentary	a.	Hair
___ 2.	Skeletal	b.	Spinal cord
___ 3.	Muscular	c.	Hormones
___ 4.	Nervous	d.	Tendons
___ 5.	Endocrine	e.	Joints

Group B

___ 6.	Circulatory	a.	Pharynx
___ 7.	Lymphatic	b.	Ureters
___ 8.	Urinary	c.	Larynx
___ 9.	Digestive	d.	Genitalia
___ 10.	Respiratory	e.	Spleen
___ 11.	Reproductive	f.	Capillaries

Circle the one that does <u>*not*</u> *belong.*

12. Pharynx	Trachea	Mouth	Alveoli
13. Uterus	Rectum	Gonads	Prostate
14. Veins	Arteries	Heart	Pancreas
15. Pineal	Bladder	Ureters	Urethra
16. Tendon	Smooth	Joints	Voluntary
17. Pituitary	Brain	Spinal cord	Nerves
18. Cartilage	Joints	Ligaments	Tendons
19. Hormones	Pituitary	Pancreas	Appendix
20. Thymus	Nails	Hair	Oil glands
21. Esophagus	Pharynx	Mouth	Trachea
22. Thymus	Spleen	Tonsils	Liver

Fill in the missing area.

SYSTEM	ORGANS	FUNCTIONS
23. Integumentary	Skin, nails, hair, sense receptors, sweat glands, oil glands	protect body from bacteria
24. Skeletal	Bones, cartilage ligaments	Support, movement, storage of minerals, blood formation
25. Muscular	Muscles	
26. Nervous	Brain, spinal cord, nerves	Communication, integration, control, recognition of sensory stimuli
27. Endocrine	pituatary, thyroid	Secretion of hormones; communication, integration, control
28. Circulatory	Heart, blood vessels	
29. Lymphatic		Transportation, immune system
30. Urinary	Kidneys, ureters, bladder, urethra	Elimination of wastes, electrolyte balance, acid-base balance, water balance
31. Digestive	Stomach, esophagus intestines	Digestion of food, absorption of nutrients
32. Respitory	Nose, pharynx, larynx, trachea, bronchi, lungs	Exchange of gases in the lungs
33. Reproductive	gonads, testes fallopian tubes ovaries	Survival of species; production of sex cells, fertilization, development, birth; nourishment of offspring; production of hormones

▸ If you have had difficulty with this section, review pages 48-59. ◂

Unscramble the words.

34. RTAHE

35. IEPLNA

36. EENVR

37. SUHESOPGA

Take the circled letters, unscramble them, and fill in the statement.

The more thoroughly you review this chapter the less

38. ☐☐☐☐☐☐☐ **you will be during your test**

APPLYING WHAT YOU KNOW

39. Myrna was 15 years old and had not yet started menstruating. Her family physician decided to consult two other physicians, each of whom specialized in a different system. Specialists in the areas of ____________________ and ____________________ were consulted.

40. Brian was admitted to the hospital with second-degree and third-degree burns over 50% of his body. He was placed in isolation, so when Jenny went to visit him, she was required to wear a hospital gown and mask. Why was Brian placed in isolation? Why was Jenny required to wear special attire?

DID YOU KNOW?

Muscles comprise 40% of your body weight. Your skeleton, however, only accounts for 18% of your body weight.

ORGAN SYSTEMS

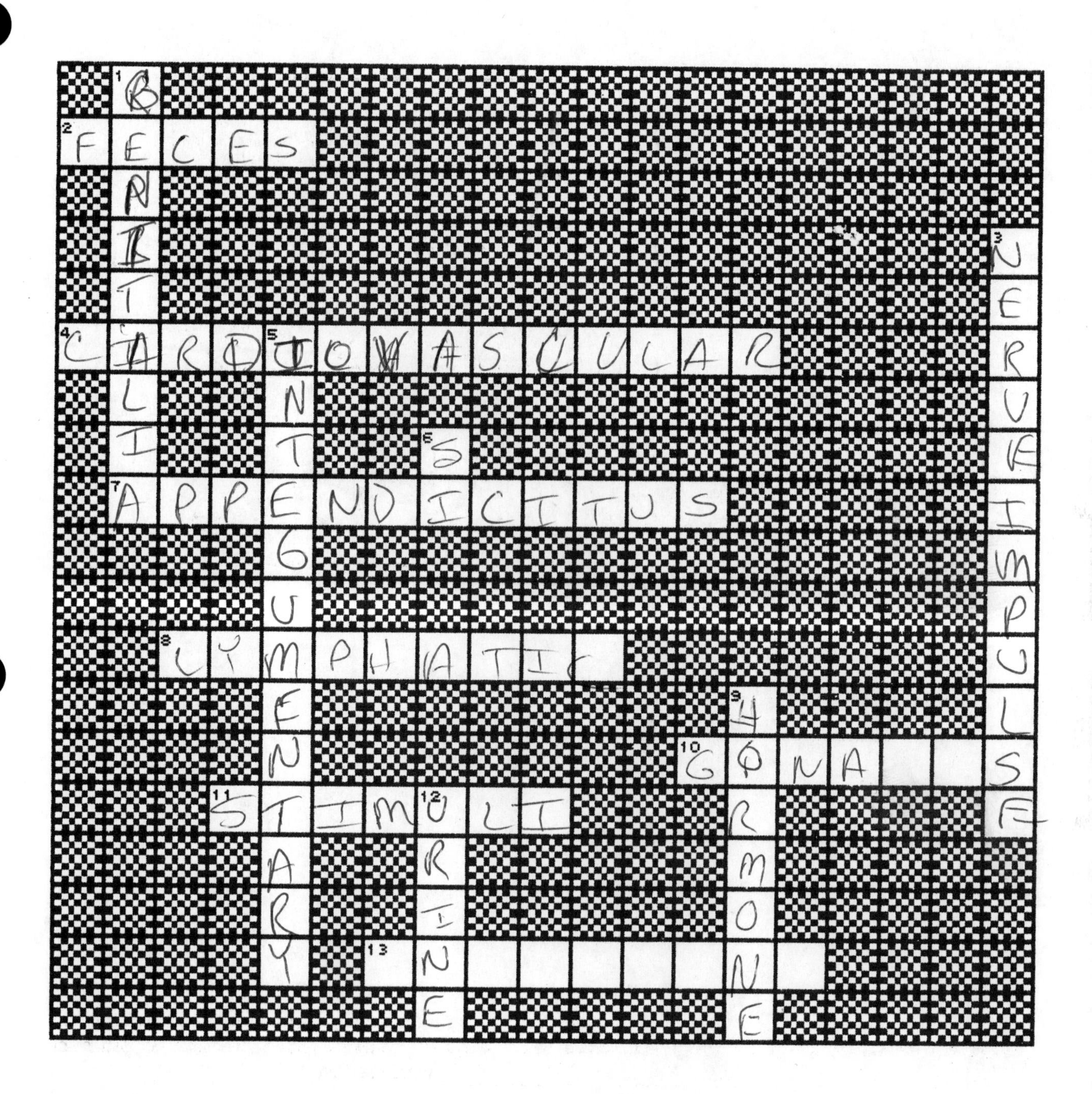

ACROSS

2. Undigested residue of digestion
4. Heart and blood vessels
7. Inflammation of the appendix
8. Subdivision of circulatory system
10. Testes and ovaries
11. Agent that causes change in the activity of a structure

DOWN

1. Vulva, penis, and scrotum
3. Specialized signal of nervous system (two words)
5. Skin
6. Gastrointestinal tract (abbrev.)
9. Chemical secretion of endocrine system
12. Waste product of kidneys

CHAPTER 4

The Integumentary System and Body Membranes

More of our time, attention, and money are spent on this system than any other one. Every time we look into a mirror we become aware of the integumentary system, as we observe our skin, hair, nails, and the appendages that give luster and comfort to this system. The discussion of the skin begins with the structure and function of the two primary layers called the epidermis and dermis. It continues with an examination of the appendages of the skin, which include the hair, receptors, nails, sebaceous glands, and sudoriferous glands. Your study of skin concludes with a review of one of the most serious and frequent threats to the skin — burns. An understanding of the integumentary system provides you with an appreciation of the danger that severe burns could pose to this system.

Membranes are thin sheetlike structures that cover, protect, anchor, or lubricate body surfaces, cavities, or organs. The two major categories are epithelial and connective. Each type is located in specific areas of the body and is vulnerable to specific disease conditions. Knowledge of the location and function of these membranes prepares you for the study of their relationship to other systems and the body as a whole.

TOPICS FOR REVIEW

Before progressing to Chapter 5, you should have an understanding of the skin and its appendages. Your review should include the classification of burns and the method used to estimate the percentage of body surface area affected by burns. A knowledge of the types of body membranes, their location, and their function is necessary to complete your study of this chapter.

CLASSIFICATION OF BODY MEMBRANES

Select the best answer.

(a) Cutaneous (b) Serous (c) Mucous (d) Synovial

___ 1. Pleura
___ 2. Lines joint spaces
___ 3. Respiratory tract
___ 4. Skin
___ 5. Peritoneum
___ 6. Contains no epithelium
___ 7. Urinary tract
___ 8. Lines body surfaces that open directly to the exterior

▸ If you have had difficulty with this section, review pages 64-66. ◂

THE SKIN

Match the term on the left with the proper selection on the right.

Group A

d 9. Integumentary system
a 10. Epidermis
b 11. Dermis
c 12. Subcutaneous
e 13. Cutaneous membrane

a. Outermost layer of skin
b. Deeper of the two layers of skin
c. Allows for rapid absorption of injected material
d. The skin is the primary organ
e. Composed of dermis and epidermis

Group B

a 14. Keratin
d 15. Melanin
e 16. Stratum corneum
c 17. Dermal papillae
b 18. Cyanosis

a. Protective protein
b. Blue-gray color of skin resulting from a decrease in oxygen
c. Rows of peglike projections
d. Brown pigment
e. Outer layer of epidermis

Select the best answer.

(a) Epidermis (b) Dermis

____ 19. Tightly packed epithelial cells
b 20. Nerves
____ 21. Fingerprints
a 22. Blisters
____ 23. Keratin
b 24. Connective tissue
____ 25. Follicle
b 26. Sebaceous gland
____ 27. Sweat gland
a 28. More cellular than other layer

▸ If you have had difficulty with this section, review pages 66-73. ◂

Fill in the blanks.

29. The three most important functions of the skin are protection, temp regulator and sense organ activity

30. Keratin prevents the sun's ultraviolet rays from penetrating the interior of the body.

31. The hair of a newborn infant is called lanugo.

32. Hair growth begins from a small cap-shaped cluster of cells called the hair papilla.

33. Depilatories act by dissolving the protein in hair shafts that extend above the skin surface.

34. The ______________ muscle produces "goose pimples."

35. Meissner's corpuscle is generally located rather close to the skin surface and is capable of detecting sensations of ______________.

36. The most numerous, important, and widespread sweat glands in the body are the ______________ sweat glands.

37. The ______________ sweat glands are found primarily in the axilla and in the pigmented skin areas around the genitals.

38. ______________ has been described as "nature's skin cream."

Circle the correct answer.

39. A first-degree burn (will or will not) blister.

40. A second-degree burn (will or will not) scar.

41. A third-degree burn (will or will not) have pain immediately.

42. According to the "rule of nines" the body is divided into (9 or 11) areas of 9%.

43. Destruction of the subcutaneous layer occurs in (second- or third-) degree burns.

▸ If you have had difficulty with this section, review pages 70-75. ◂

Unscramble the words.

44. PIDEEMIRS

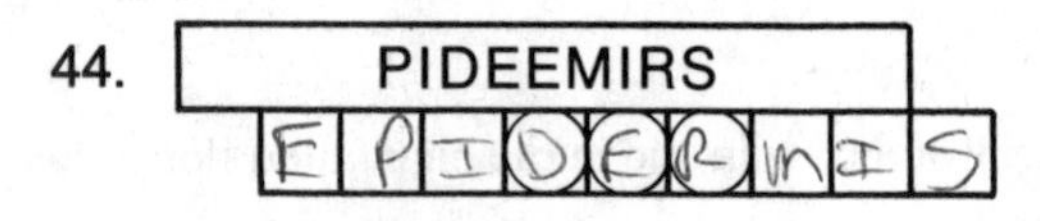

45. REKTAIN

46. AHIR

47. UGONAL

48. DRTONIDEHYA

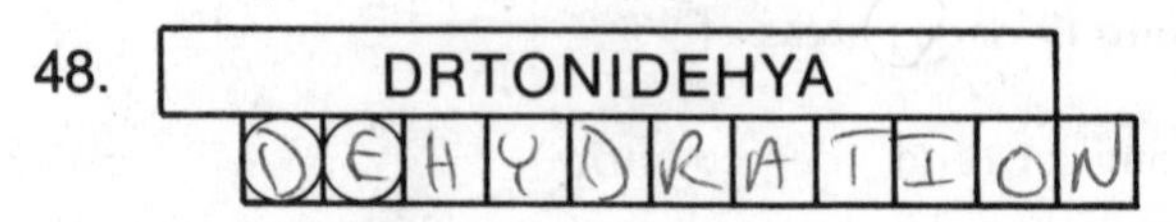

Take the circled letters, unscramble them, and fill in the statement.

What Amanda's mother gave her after every date.

49.

APPLYING WHAT YOU KNOW

50. Mr. Ziven was admitted to the hospital with second-degree and third-degree burns. Both arms, the anterior trunk, the right anterior leg, and the genital region were affected by the burns. The doctor quickly estimated that ______% of Mr. Ziven's body had been burned.

51. Mrs. James complained to her doctor that she had severe pain in her chest and feared that she was having a heart attack. An ECG revealed nothing unusual, but Mrs. James insisted that every time she took a breath she experienced pain. What might be the cause of Mrs. James' pain?

DID YOU KNOW?

Because the dead cells of the epidermis are constantly being worn and washed away, we get a new outer skin layer every 27 days.

SKIN/BODY MEMBRANES

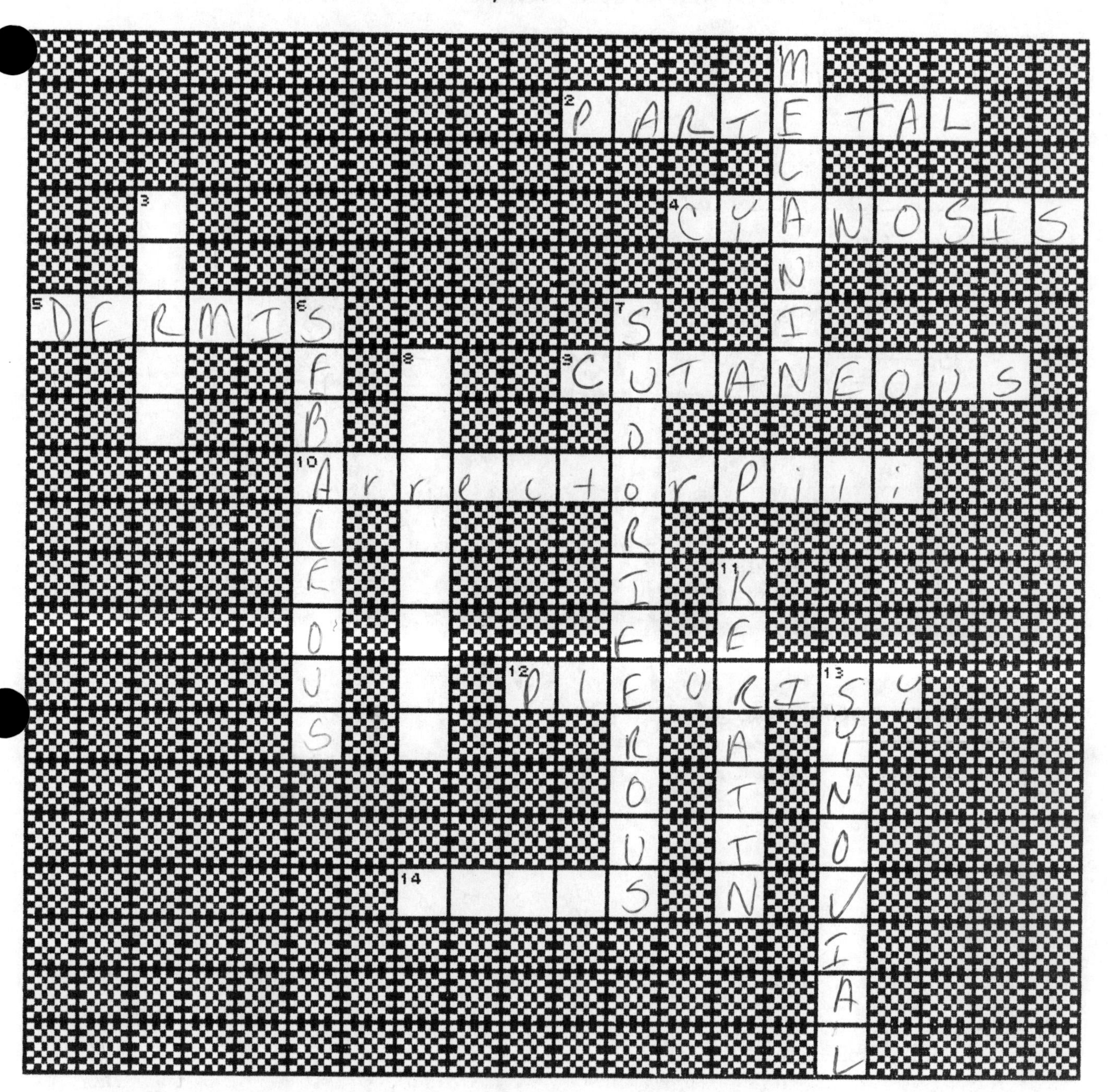

ACROSS

2. Covers the surface of organs found in serous body cavities
4. Bluish gray color of skin due to decreased 02
5. Deeper of the two primary skin layers
9. Skin
10. "Goose pimples" (two words)
12. Inflammation of the serous membrane that lines the chest and covers the lungs

DOWN

1. Brown pigment
3. Cushionlike sacs found between moving body parts
6. Oil gland
7. Sweat gland
8. Forms the lining of serous body cavities
11. Tough waterproof substance that protects body from excessive fluid loss
13. Membrane that lines joint spaces

LONGITUDINAL SECTION OF THE SKIN

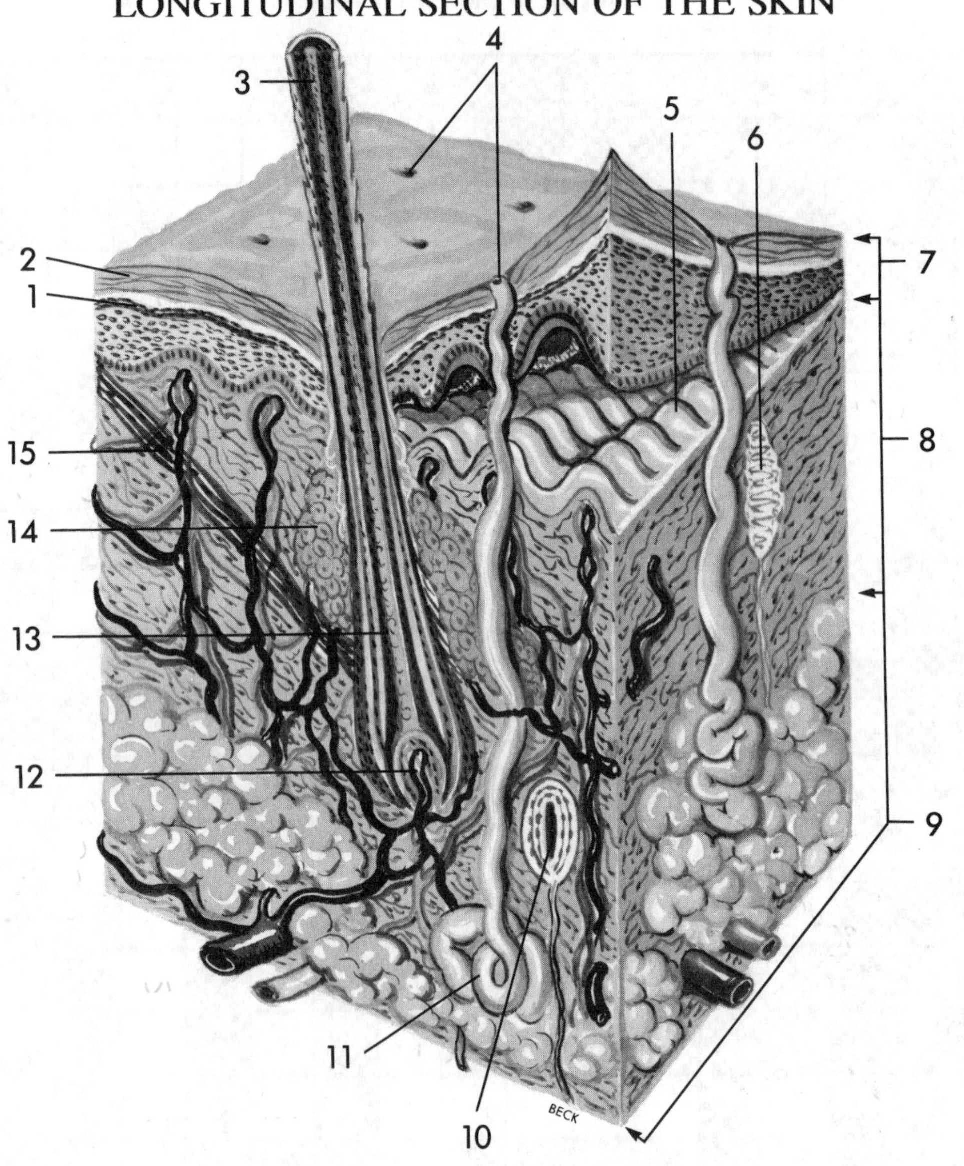

1. Pigment Layer
2. Stratum coreum
3.
4. Openings of Sweat ducts
5.
6.
7.
8.
9.
10.
11.
12.
13.
14.
15.

"RULE OF NINES"
FOR ESTIMATING SKIN SURFACE BURNED

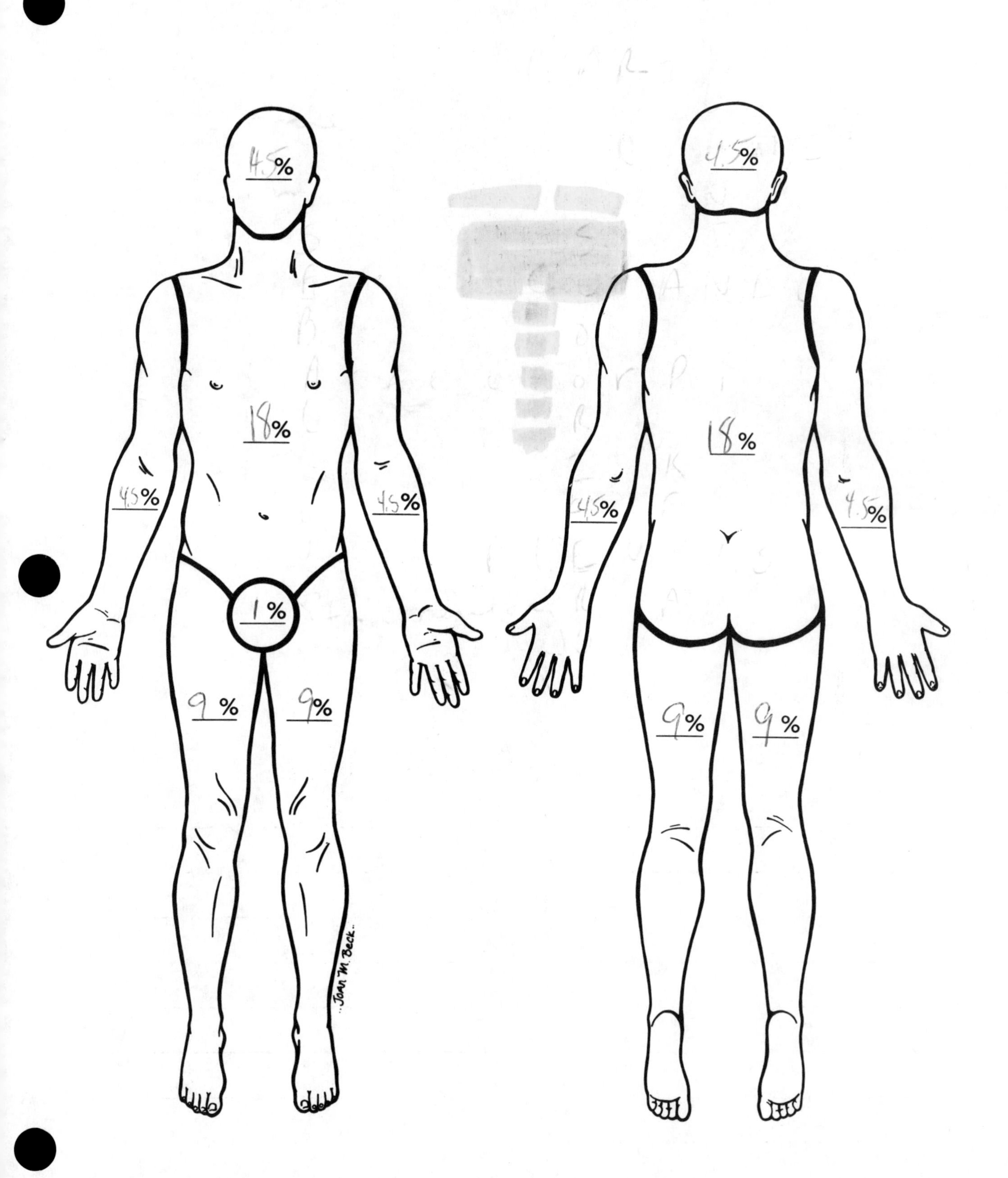

CHAPTER 5

The Skeletal System

How strange we would look without the skeleton, because it is the skeleton that provides us with the rigid, supportive framework that gives shape to our bodies. But this is just the beginning, since it also protects the organs beneath it, maintains homeostasis of blood calcium, produces blood cells, and assists the muscular system in providing movement for us.

After reviewing the microscopic structure of bone and cartilage, you will understand how skeletal tissues are formed, their differences, and their importance in the human body. Your microscopic investigation will make the study of this system easier as you logically progress from this view to macroscopic bone formation and growth and visualize the structure of long bones.

The skeleton is divided into two main divisions: the axial skeleton and the appendicular skeleton. All of the 206 bones of the human body may be classified into one of these two areas. And, although we can divide them neatly by this system, we are still aware that subtle differences exist between a man's and a woman's skeleton. These structural differences provide us with insight to the differences in function between men and women.

Finally, three types of joints exist in the body. They are synarthrosis, amphiarthrosis, and diarthrosis. It is important to have a knowledge of these joints to understand how movement is facilitated by articulations.

TOPICS FOR REVIEW

Before progressing to Chapter 6, you should familiarize yourself with the functions of the skeletal system, the structure and function of bone and cartilage, bone formation and growth, and the types of joints found in the body. Additionally, your understanding of the skeletal system should include identification of the two major subdivisions of the skeleton, the bones found in each area, and any differences that exist between a man's and a woman's skeleton.

BONE FORMATION AND GROWTH

If the statement is true, write T in the answer blank. If the statement is false, correct the statement by circling the incorrect term and inserting the correct term in the answer blank.

____T____ 1. When the skeleton forms in a baby before birth, it consists of cartilage and fibrous structures.
____F____ 2. The diaphyses are the ends of the bone. epiphyses
____T____ 3. Bone-forming cells are known as osteoclasts.
_________ 4. It is the combined action of osteoblasts and osteoclasts that sculpts bones into their adult shapes.
____T____ 5. The stresses placed on certain bones during exercise decrease the rate of bone deposition.
____F____ 6. The epiphyseal plate can be seen in both external and cutaway views of an adult long bone.
____F____ 7. The shaft of a long bone is known as the articulation.
_________ 8. Cartilage in the newborn becomes bone when it is replaced with calcified bone matrix deposited by osteoblasts.
_________ 9. When epiphyseal cartilage becomes bone, growth begins.
_________ 10. The epiphyseal cartilage is visible, if present, on x-ray films.

▸ If you have had difficulty with this section, review page 80. ◂

MICROSCOPIC STRUCTURE OF BONE AND CARTILAGE

Match the term on the left with the proper selection on the right.

Group A

__d__ 11. Trabeculae
__b__ 12. Compact
__e__ 13. Spongy
__a__ 14. Periosteum
__c__ 15. Cartilage

a. Outer covering of bone
b. Dense bone tissue
c. Fibers embedded in a firm gel
d. Needlelike threads of spongy bone
e. Ends of long bones

Group B

__D__ 16. Osteocytes
__A__ 17. Canaliculi
__E__ 18. Lamellae
__B__ 19. Chondrocytes
__C__ 20. Haversian system

a. Connect lacunae
b. Cartilage cells
c. Structural unit of compact bone
d. Bone cells
e. Ring of bone

▸ If you have had difficulty with this section, review pages 80-83. ◂

TYPES OF BONES
STRUCTURE OF LONG BONES
FUNCTIONS

Fill in the blanks.

21. There are __four__ types of bones.

22. The ______________ ______________ is the hollow area inside the diaphysis of a bone.

23. A thin layer of cartilage covering each epiphysis is the ______________.

24. The __endosteum__ lines the medullary cavity of long bones.

25. __Hemopoisis__ is used to describe the process of blood cell formation.

Circle the correct choice.

26. Which one of the following is *not* a part of the axial skeleton?

a. Scapula
b. Cranial bones
c. Vertebra
d. Ribs
e. Sternum

27. Which one of the following is *not* a cranial bone?

a. Frontal
b. Parietal
c. Occipital
d. Lacrimal
e. Sphenoid

28. Which of the following is *not* correct?

a. A baby is born with a straight spine.
b. In the adult the sacral and thoracic curves are convex.
c. The normal curves of the adult spine provide greater strength than a straight spine.
d. A curved structure has more strength than a straight one of the same size and materials.

29. True ribs:

a. Attach to the cartilage of other ribs
b. Do not attach to the sternum
c. Attach directly to the sternum without cartilage
d. Attach directly to the sternum by means of cartilage

30. The bone that runs along the thumb side of your forearm is the:

a. Humerus
b. Ulna
c. Radius
d. Tibia

31. The shinbone is also known as the:

a. Fibula
b. Femur
c. Tibia
d. Ulna

32. The bones in the palm of the hand are called:

a. Metatarsals
b. Tarsals
c. Carpals
d. Metacarpals

33. Which one of the following is *not* a bone of the upper extremity?

a. Radius
b. Clavicle
c. Humerus
d. Ilium

34. The heel bone is known as the:

a. Calcaneus
b. Talus
c. Metatarsal
d. Phalanges

35. The mastoid process is part of which bone?

a. Parietal
b. Temporal
c. Occipital
d. Frontal

36. When a baby learns to walk, which area of the spine becomes concave?

a. Lumbar
b. Thoracic
c. Cervical
d. Coccyx

37. Which bone is the "funny" bone?

a. Radius
b. Ulna
c. Humerus
d. Carpal

38. There are how many pair of true ribs?

a. 14
b. 7
c. 5
d. 3

39. The 27 bones in the wrist and the hand allow for more:

a. Strength
b. Dexterity
c. Protection
d. Red blood cell products

40. The longest bone in the body is the:

a. Tibia
b. Fibula
c. Femur
d. Humerus

41. Distally, the _______________ articulates with the patella.

a. Femur
b. Fibula
c. Tibia
d. Humerus

42. These bones form the cheek bones:

a. Mandible
b. Palatine
c. Maxillary
d. Zygomatic

43. In a child, there are five of these bones. In an adult, they are fused into one:

a. Pelvic
b. Lumbar vertebrae
c. Sacrum
d. Carpals

44. The spinal cord enters the cranium through a large hole (foramen magnum) in this bone:

a. Temporal
b. Parietal
c. Occipital
d. Sphenoid

Circle the one that does not belong.

45. Cervical	Thoracic	Os coxae	Coccyx
46. Pelvic girdle	Ankle	Wrist	Axial
47. Frontal	Occipital	Maxillary	Sphenoid
48. Scapula	Pectoral girdle	Ribs	Clavicle
49. Malleus	Vomer	Incus	Stapes
50. Ulna	Ilium	Ischium	Pubis

51. Carpal	Phalanges	Metacarpal	Ethmoid
52. Ethmoid	Parietal	Occipital	Nasal
53. Anvil	Atlas	Axis	Cervical

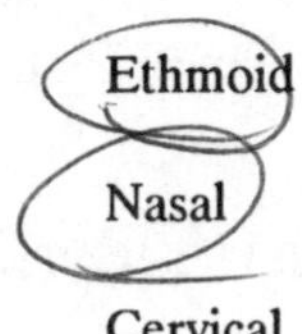

▸ If you have had difficulty with this section, review pages 85-99. ◂

DIFFERENCES BETWEEN A MAN'S AND A WOMAN'S SKELETON

Choose the right answer.

(a) Male (b) Female

__a__ 54. Funnel-shaped pelvis
__b__ 55. Broader-shaped pelvis
__b__ 56. Osteoporosis occurs more frequently
__a__ 57. Larger
__b__ 58. Wider pelvic brim

▸ If you have had difficulty with this section, review pages 84 and 99. ◂

BONE MARKINGS

From the choices given, match the bone with its identification marking. Bones may be used more than once.

a. Mastoid
b. Pterygoid process
c. Foramen magnum
d. Sella turcica
e. Mental foramen
f. Conchae
g. Xiphoid process
h. Glenoid cavity
i. Olecranon process
j. Ischium
k. Acetabulum
l. Symphysis pubis
m. Ilium
n. Greater trochanter
o. Medial malleolus
p. Calcaneus
q. Acromion process
r. Frontal sinuses
s. Condyloid process
t. Tibial tuberosity

________ 59. Occipital
________ 60. Sternum
________ 61. Os coxae
________ 62. Femur
________ 63. Ulna
________ 64. Temporal
________ 65. Tarsals
________ 66. Sphenoid
________ 67. Ethmoid
________ 68. Scapula
________ 69. Tibia
________ 70. Frontal
________ 71. Mandible

▸ If you have had difficulty with this section, review pages 88-99. ◂

JOINTS (ARTICULATIONS)

Circle the correct answer.

72. Freely movable joints are (amphiarthroses or diarthroses).

73. The sutures in the skull are (synarthrotic or amphiarthrotic) joints.

74. All (diarthrotic or amphiarthrotic) joints have a joint capsule, a joint cavity, and a layer of cartilage over the ends of the two joining bones.

75. (Ligaments or tendons) grow out of periosteum and attach two bones together.

76. The (articular cartilage or epiphyseal cartilage) absorbs jolts.

77. Gliding joints are the (least movable or most movable) of the diarthrotic joints.

78. The knee is the (largest or smallest) joint.

79. Hinge joints allow motion in (2 or 4) directions.

80. The saddle joint at the base of each of our thumbs allows for greater (strength or mobility).

81. When you rotate your head, you are using a (gliding or pivot) joint.

▸ If you have had difficulty with this section, review pages 101-104. ◂

Unscramble the bones.

82. ETVERRBAE

83. BPSUI

84. SCALUPA

85. IMDBALNE

86. APNHGAELS

Take the circled letters, unscramble them, and fill in the statement.

What the fat lady wore to the ball.

87.

APPLYING WHAT YOU KNOW

88. Mrs. Perine had advanced cancer of the bone. As the disease progressed, Mrs. Perine required several blood transfusions throughout her therapy. She asked the doctor one day to explain the necessity for the transfusions. What explanation might the doctor give to Mrs. Perine?

89. Dr. Kennedy, an orthopedic surgeon, called the admissions office of the hospital and advised that he would be admitting a patient in the next hour with an epiphyseal fracture. Without any other information, the patient is assigned to the pediatric ward. What prompted this assignment?

90. Mrs. Van Skiver, age 60, noticed when she went in for her physical examination that she was a half inch shorter than she was on her last visit. Dr. Veazey suggested she begin a regimen of dietary supplements of calcium, vitamin D, and a prescription for sex hormone therapy. What bone disease did Dr. Veazey suspect?

DID YOU KNOW?

The bones of the hands and feet make up more than half of the total 206 bones of the body.

SKELETAL SYSTEM

ACROSS

2. Lines joint capsule
3. Cartilage cells
5. Joint
7. Space inside cranial bone
9. Bone cells
10. Type of bone
11. Division of skeleton
12. Freely movable joints
13. Ends of long bones

DOWN

1. Process of blood cell formation
4. Covers long bone except at its joint surfaces
6. Bone absorbing cells
7. Suture joints
8. Spaces in bones where osteocytes are found

LONGITUDINAL SECTION OF LONG BONE

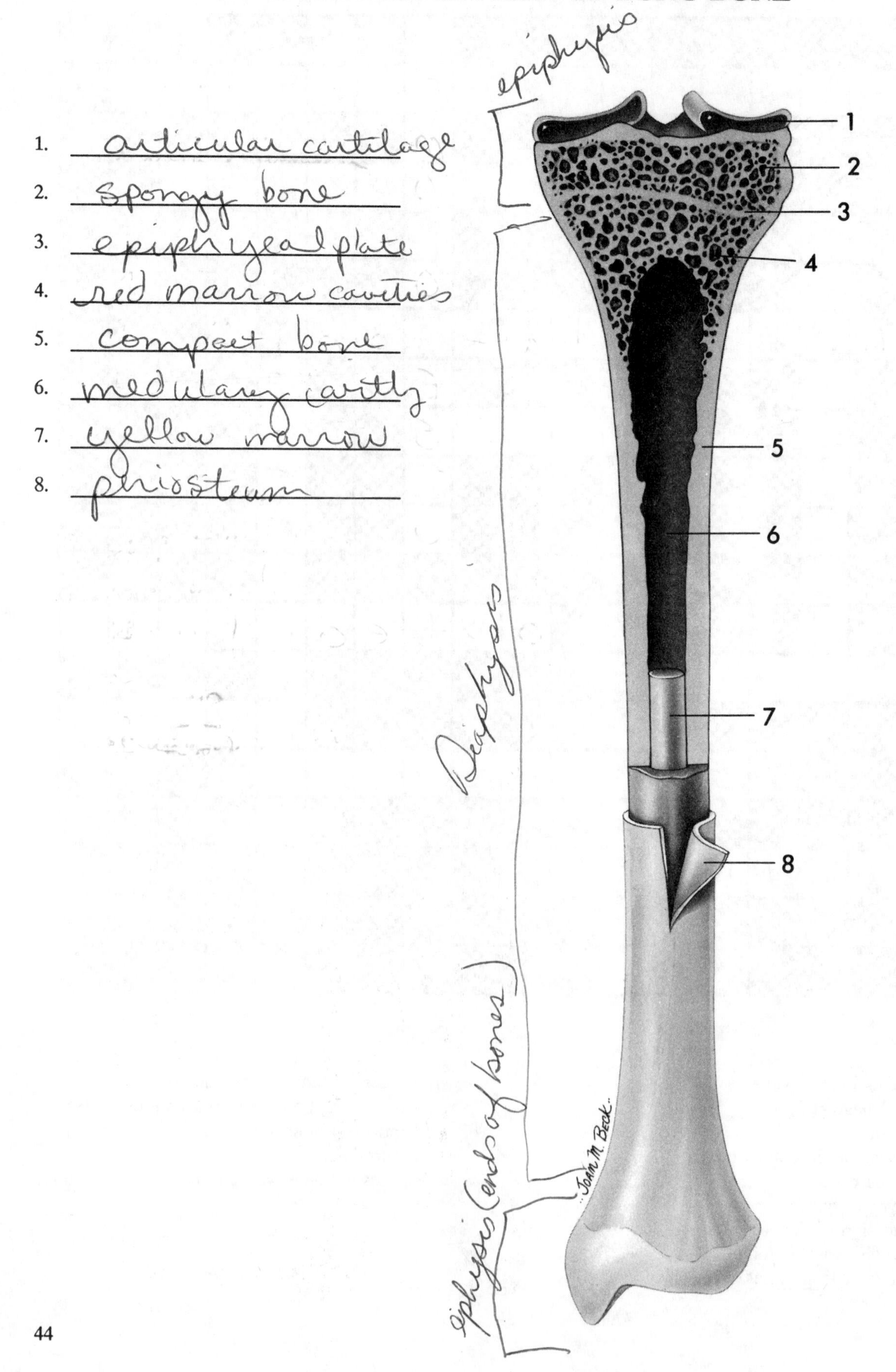

ANTERIOR VIEW OF SKELETON

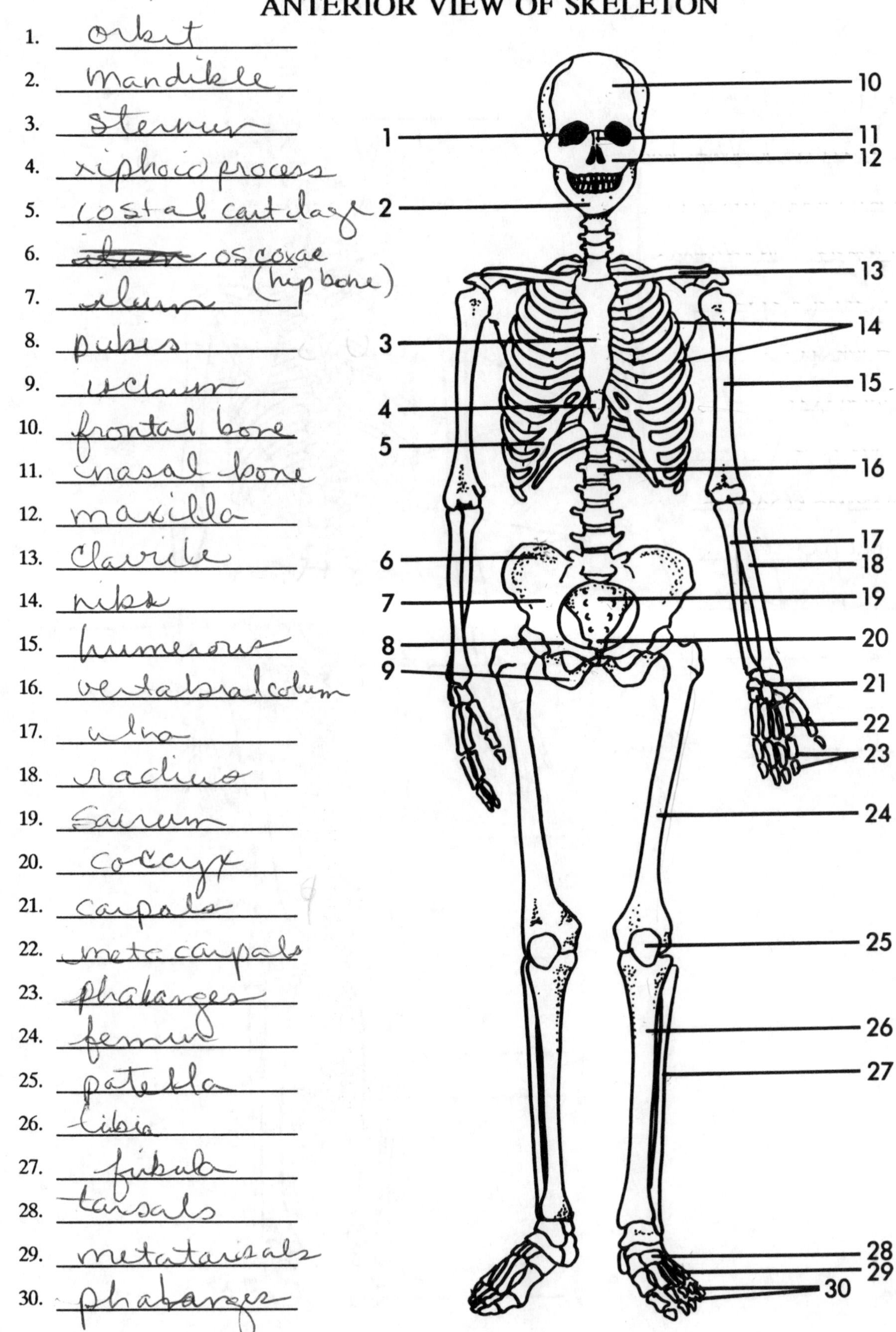

POSTERIOR VIEW OF SKELETON

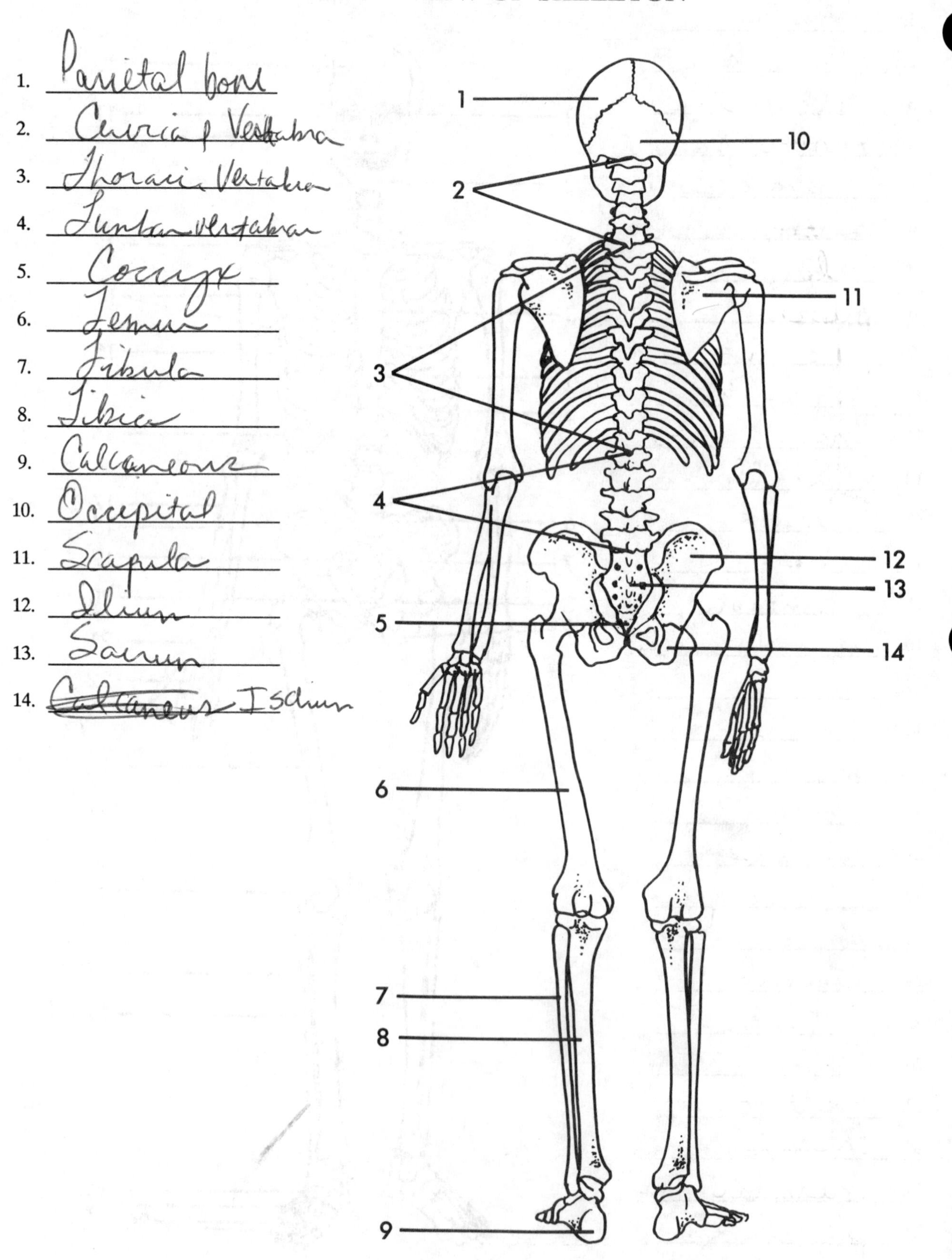

SKULL VIEWED FROM THE RIGHT SIDE

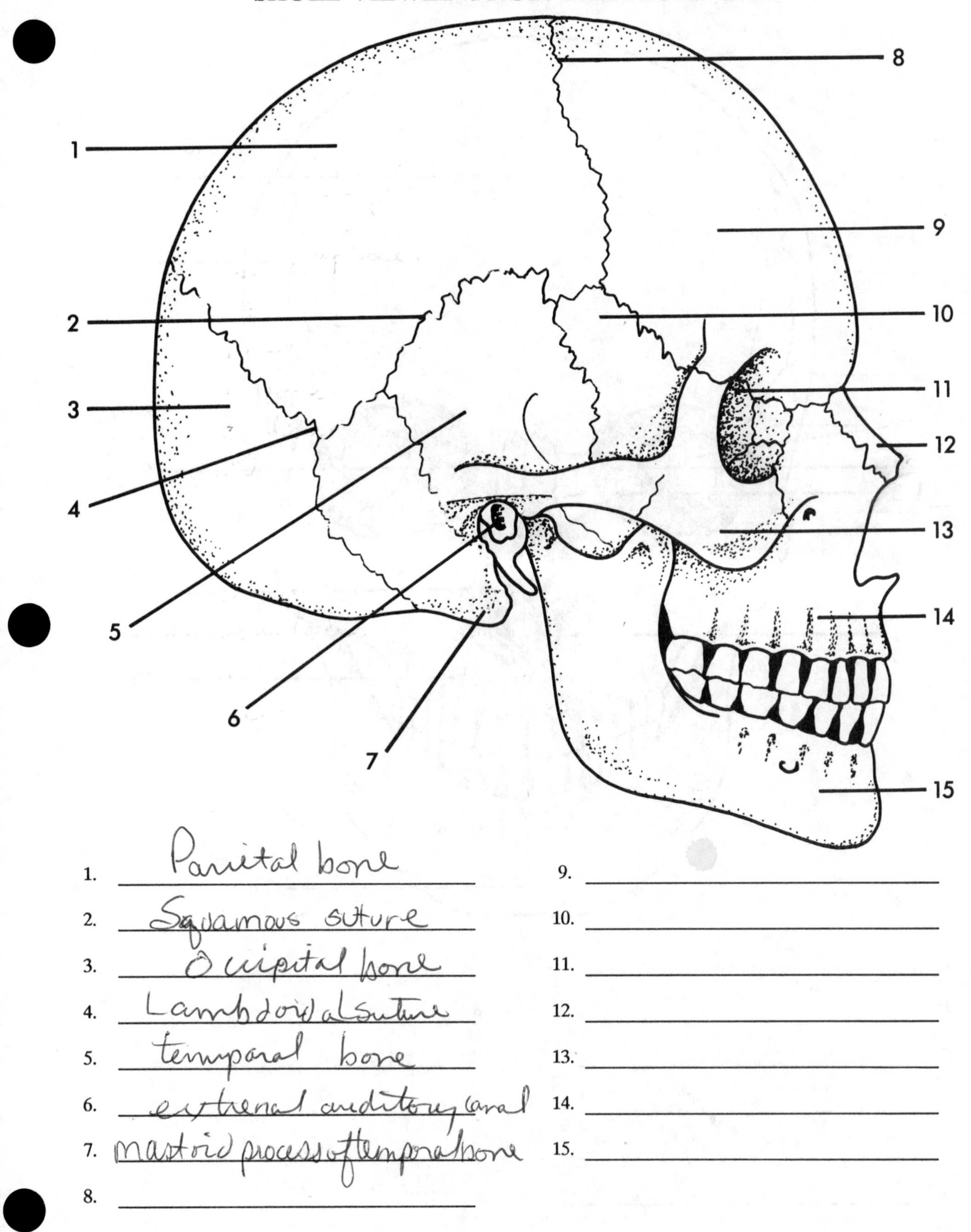

1. Parietal bone
2. Squamous suture
3. Occipital bone
4. Lambdoidal suture
5. temporal bone
6. external auditory canal
7. mastoid process of temporal bone
8. __________
9. __________
10. __________
11. __________
12. __________
13. __________
14. __________
15. __________

SKULL VIEWED FROM THE FRONT

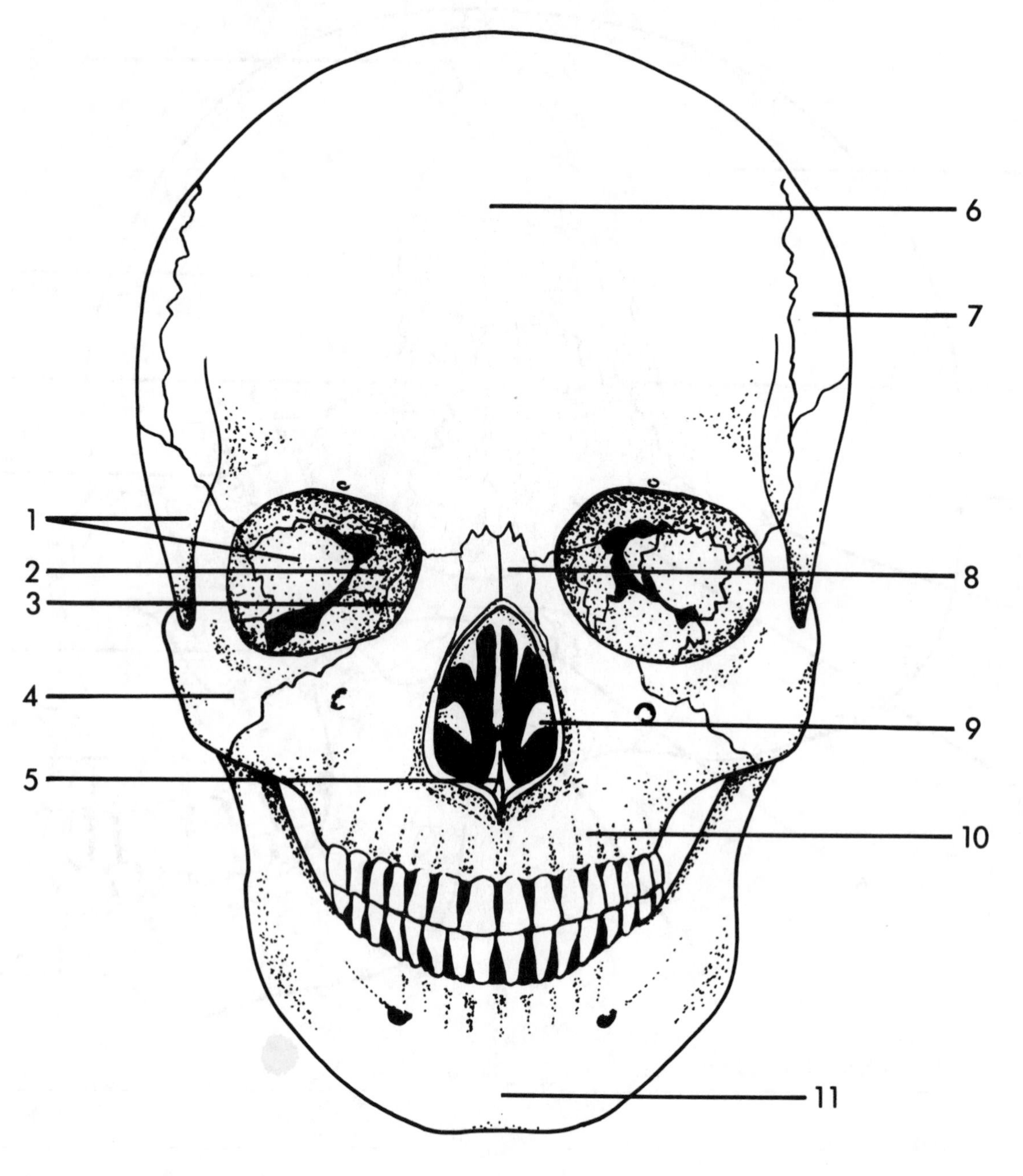

1. ______________________
2. ______________________
3. ______________________
4. ______________________
5. ______________________
6. ______________________
7. ______________________
8. ______________________
9. ______________________
10. ______________________
11. ______________________

STRUCTURE OF A DIARTHROTIC JOINT

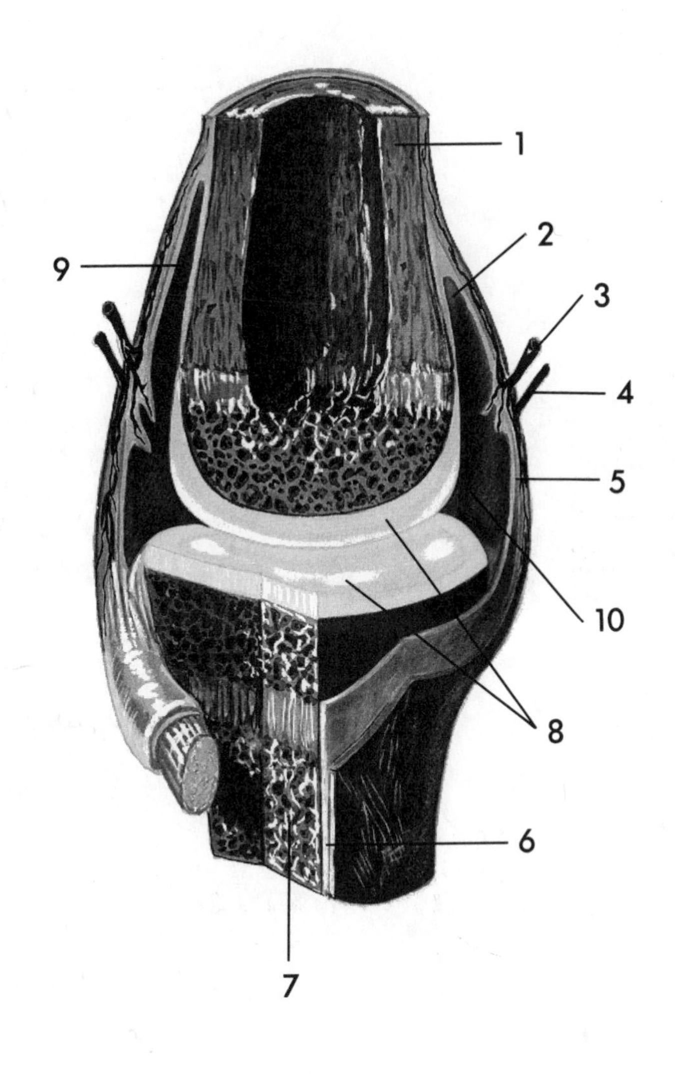

1. ______________________
2. ______________________
3. ______________________
4. ______________________
5. ______________________
6. ______________________
7. ______________________
8. ______________________
9. ______________________
10. ______________________

CHAPTER 6

The Muscular System

The muscular system is often referred to as the "power system," and rightfully so, because it is this system that provides the motion necessary to move the body and perform organic functions. Just as an automobile relies on the engine to provide motion, the body depends on the muscular system to perform both voluntary and involuntary types of movement. Walking, breathing, and the digestion of food are but a few examples of body functions that require the healthy performance of the muscular system.

Although this system has several functions, the primary purpose is to provide movement or power. Muscles produce power by contracting. The ability of a large muscle or muscle group to contract depends on the ability of microscopic muscle fibers that contract within the larger muscle. An understanding of these microscopic muscle fibers will assist you as you progress in your study to the larger muscles and muscle groups.

Muscle contractions may be one of several types: isotonic, isometric, twitch, or tetanic. When skeletal or voluntary muscles contract, they provide us with a variety of motions. Flexion, extension, abduction, adduction, and rotation are examples of these movements that provide us with both strength and agility.

Muscles must be used to keep the body healthy and in good condition. Scientific evidence keeps pointing to the fact that the proper use and exercise of muscles may prolong longevity. An understanding of the structure and function of the muscular system may, therefore, add quality and quantity to our lives.

TOPICS FOR REVIEW

Before progressing to Chapter 7, you should familiarize yourself with the structure and function of the three major types of muscle tissue. Your review should include the microscopic structure of skeletal muscle tissue, how a muscle is stimulated, the major types of skeletal muscle contractions, and the skeletal muscle groups. Your study should conclude with an understanding of the types of movements produced by skeletal muscle contractions.

MUSCLE TISSUE

Select the correct term from the choices given and insert the letter in the answer blank.

(a) Skeletal muscle (b) Cardiac muscle (c) Smooth muscle

____ 1. Striated
____ 2. Cells branch frequently
____ 3. Moves food into stomach
____ 4. Nonstriated
____ 5. Voluntary
____ 6. Keeps blood circulating through its vessels
____ 7. Involuntary
____ 8. Attaches to bone
____ 9. Hollow internal organs
____ 10. Maintenance of normal blood pressure

▸ If you have had difficulty with this section, review page 108. ◂

SKELETAL MUSCLES

Match the term on the left with the proper selection on the right.

Group A

_____ 11. Origin
_____ 12. Insertion
_____ 13. Body
_____ 14. Tendons
_____ 15. Bursae

a. The muscle unit excluding the ends
b. Attachment to the more movable bone
c. Fluid-filled sacs
d. Attachment to more stationary bone
e. Attach muscle to bones

MICROSCOPIC STRUCTURE

Group B

_____ 16. Muscle fibers
_____ 17. Actin
_____ 18. Sarcomere
_____ 19. Myosin
_____ 20. Myofilament

a. Protein that forms thick myofilaments
b. Basic functional unit of skeletal muscle
c. Protein that forms thin myofilaments
d. Microscopic threadlike structures found in skeletal muscle fibers
e. Specialized contractile cells of muscle tissue

▸ If you have had difficulty with this section, review pages 108-111. ◂

FUNCTIONS

Fill in the blanks.

21. Muscles move bones by _____________ on them.
22. As a rule, only the _____________ bone moves.
23. The _____________ bone moves toward the _____________ bone.
24. Of all the muscles contracting simultaneously, the one mainly responsible for producing a particular movement is called the _____________ for that movement.
25. As prime movers contract, other muscles called _____________ relax.
26. The biceps brachii is the prime mover during flexing, and the brachialis is its helper or _____________ muscle.
27. We are able to maintain our body position because of a specialized type of skeletal muscle contraction called _____________
28. _____________ maintains body posture by counteracting the pull of gravity.
29. A decrease in temperature, a condition know as _____________, will drastically affect cellular activity and normal body function.

30. Energy required to produce a muscle contraction is obtained from ATP.

▸ If you have had difficulty with this section, review pages 111-112. ◂

FATIGUE
MOTOR UNIT
MUSCLE STIMULUS

*If the statement is true, write **T** on the answer blank. If the statement is false, correct the statement by circling the incorrect term and inserting the correct term in the answer blank.*

_________ 31. The point of contact between the nerve ending and the muscle fiber is called a motor neuron.
_________ 32. A motor neuron together with the cells it innervates is called a motor unit.
_________ 33. If muscle cells are stimulated repeatedly without adequate periods of rest, the strength of the muscle contraction will decrease resulting in fatigue.
_________ 34. The depletion of oxygen in muscle cells during vigorous and prolonged exercise is known as fatigue.
_________ 35. An adequate stimulus will contract a muscle cell completely because of the "must" theory.
_________ 36. When oxygen supplies run low, muscle cells produce ATP and other waste products during contraction.
_________ 37. Exercise physiologists in the Soviet Union have found that the zero-gravity environment of space promotes a loss of postural drainage.
_________ 38. The minimal level of stimulation required to cause a fiber to contract is called the threshold stimulus.
_________ 39. Smooth muscles bring about movements by pulling on bones across movable joints.
___T___ 40. A nervous system disorder that shuts off impulses to certain skeletal muscles may result in paralysis.

TYPES OF SKELETAL MUSCLE CONTRACTION

Circle the correct choice.

41. When a muscle does not shorten and no movement results, the contraction is:

a. Isometric
b. Isotonic
c. Twitch
d. Tetanic

42. Walking is an example of which type of contraction?

a. Isometric
b. Isotonic
c. Twitch
d. Tetanic

43. Pushing against a wall is an example of which type of contraction?

a. Isotonic
b. Isometric
c. Twitch
d. Tetanic

44. Endurance training is also known as:

a. Isometrics
b. Hypertrophy
c. Aerobic training
d. Strength training

45. Benefits of regular exercise include all of the following <u>except</u>:

a. Improved lung functioning
b. More efficient heart
c. Less fatigue
d. Atrophy

46. Twitch contractions can be easily seen in:

a. Isolated muscles prepared for research
b. A great deal of normal muscle activity
c. During resting periods
d. None of the above

47. Individual contractions "melt" together to produce a sustained contraction or:

a. Twitch
b. Tetanus
c. Isotonic response
d. Isometric response

48. In most cases, isotonic contraction of muscle produces movement at a/an:

a. Insertion
b. Origin
c. Joint
d. Bursa

49. Prolonged inactivity causes muscles to shrink in mass, a condition called:

a. Hypertrophy
b. Disuse atrophy
c. Paralysis
d. Muscle fatigue

50. Muscle hypertrophy can be best enhanced by a program of:

a. Isotonic exercise
b. Better posture
c. High-protein diet
d. Strength training

▸ If you have had difficulty with this section, review pages 112-116. ◂

SKELETAL MUSCLE GROUPS

Choose the proper function for the muscles listed below and place the letter in the answer blank.

(a) Flexor (b) Extensor (c) Abductor (d) Adductor
(e) Rotator (f) Dorsiflexor or Plantar flexor

_____ 51. Deltoid
_____ 52. Tibialis anterior
_____ 53. Gastrocnemius
_____ 54. Biceps brachii
_____ 55. Gluteus medius
_____ 56. Soleus
_____ 57. Iliopsoas
_____ 58. Pectoralis major
_____ 59. Gluteus maximus
_____ 60. Triceps brachii
_____ 61. Sternocleidomastoid
_____ 62. Trapezius
_____ 63. Gracilis

▸ If you have had difficulty with this section, review page 117-124. ◂

TYPES OF MOVEMENTS PRODUCED BY SKELETAL MUSCLE CONTRACTIONS

Circle the correct choice.

64. A movement that makes the angle between two bones smaller is:

    a. Flexion
    b. Extension
    c. Abduction
    d. Adduction

65. Moving a part toward the midline is:

    a. Flexion
    b. Extension
    c. Abduction
    d. Adduction

66. Moving a part away from the midline is:

    a. Flexion
    b. Extension
    c. Abduction
    d. Adduction

67. When you move your head from side to side as in shaking your head "no" you are ____________________ a muscle group.

    a. Rotating
    b. Pronating
    c. Supinating
    d. Abducting

68. ____________________ occurs when you turn the palm of your hand from an anterior to posterior position.

    a. Dorsiflexion
    b. Plantar flexion
    c. Supination
    d. Pronation

69. Dorsiflexion refers to:

    a. Hand movements
    b. Eye movements
    c. Foot movements
    d. Head movements

▸ If you have had difficulty with this section, review pages 122-125. ◂

APPLYING WHAT YOU KNOW

70. Casey noticed pain whenever she reached for anything in her cupboards. Her doctor told her that the small fluid-filled sacs in her shoulder were inflamed. What condition did Casey have?

71. The nurse was preparing an injection for Mrs. Tatakis. The amount to be given was 2 ml. What area of the body will the nurse most likely select for this injection?

72. Warren was playing football and pulled a band of fibrous connective tissue that attached a muscle to a bone. What is the common term for this tissue?

73. WORD FIND

Can you find the muscles from the list below in the box of letters? Words may be spelled top to bottom, bottom to top, right to left, left to right, or diagonally.

U X D T F I N I G I R O I R G G X F Z B
K P B F L D N K Y D X S O N W R W Q N U
E I Z H M U S C L E O T O M R O E T O Z
J S R I Q E U V I T C I J G O M D E I Y
W P D R X M G S O U X J X D T R E N X G
K E N B K O D N D E U G C E A N G O E F
M C M A X I I B L N T U D T T I S S L F
S I Y M K C A F P S S T M A O B U Y F N
Z B V H P E I W L N Y E G I R U I N R O
P H S X X S B H S F H X A R F R Z O S I
G W V V R B G S O C P T R T F S E V P T
F Z W O K Q E N L O O E H S N A P I E R
A M D X E P M W E H R N P U D I A T C E
G F Y R S T O A U Q T S A Z Q H R I I S
C R Z A A N R C S E A I I E P G T S R N
U B H J H S M V N N L O D V G C T T T I
G H T S I G R E N Y S N D I O T L E D V
I S O M E T R I C Z K G E U G I T A F U
I S C Y S U I M E N C O R T S A G C Z A
T C S G N I R T S M A H R T E N D O N H

LIST OF WORDS

ORIGIN
EXTENSION
ABDUCTOR
DORSIFLEXION
STRIATED
BURSA
FATIGUE
ROTATOR
MUSCLE
TRICEPS
TRAPEZIUS
DELTOID
HAMSTRINGS
ATROPHY
TENDON
INSERTION
SOLEUS
ISOTONIC
BICEPS
SYNERGIST
DIAPHRAGM
GASTROCNEMIUS
TENOSYNOVITIS
FLEXION
ISOMETRIC

DID YOU KNOW?

If all of your muscles pulled in one direction, you would have the power to move 25 tons.

THE MUSCULAR SYSTEM

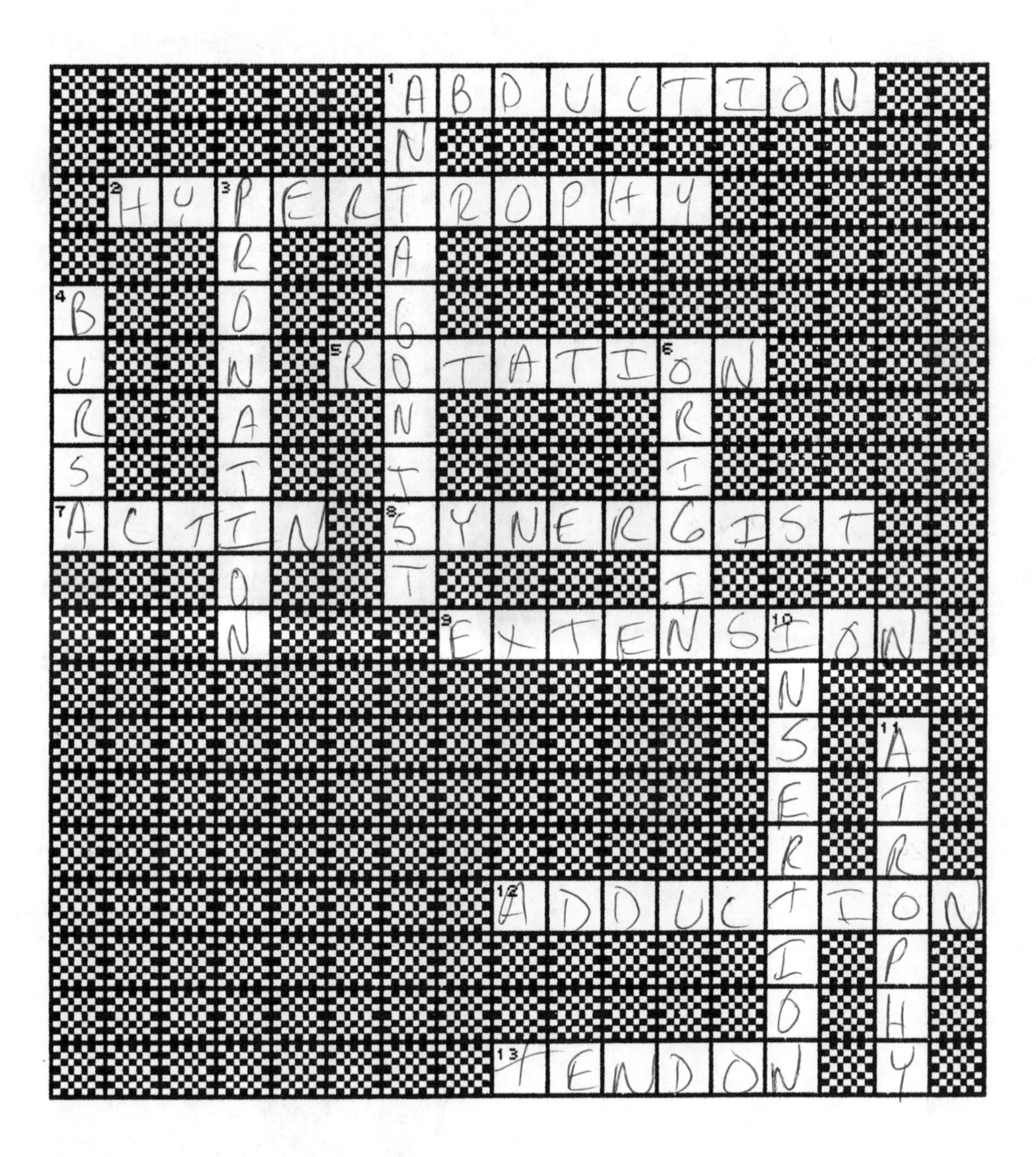

ACROSS

1. Away from the body's midline
2. Increase in size
5. Shaking your head "no"
7. Protein which composes myofilaments
8. Assist prime movers with movement
9. Movement that makes joint angles larger
12. Toward the body's midline
13. Anchors muscles to bones

DOWN

1. Produces movement opposite to prime movers
3. Turning your palm from an anterior to posterior position
4. Small fluid-filled sacs between tendons and bones
6. Attachment to the more stationary bone
10. Attachment to the more movable bone
11. Muscle shrinkage

MUSCLES—ANTERIOR VIEW

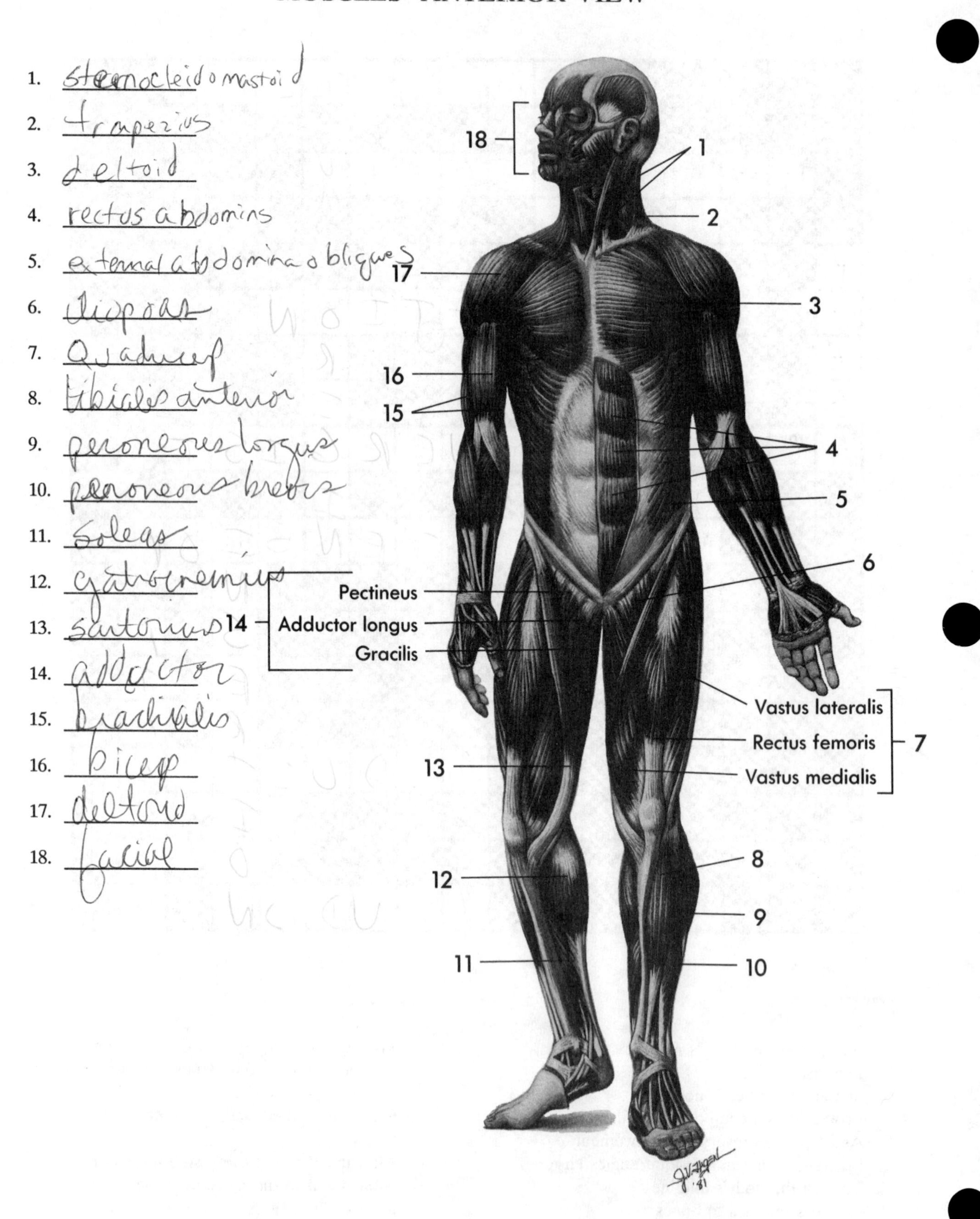

1. sternocleidomastoid
2. trapezius
3. deltoid
4. rectus abdomins
5. external abdomina obliques
6. iliopoas
7. Quadricep
8. tibialis anterior
9. peroneous longus
10. peroneous breves
11. soleas
12. gastrocnemius
13. sartorius
14. adductor
15. brachialis
16. biceps
17. deltoid
18. facial

MUSCLES—POSTERIOR VIEW

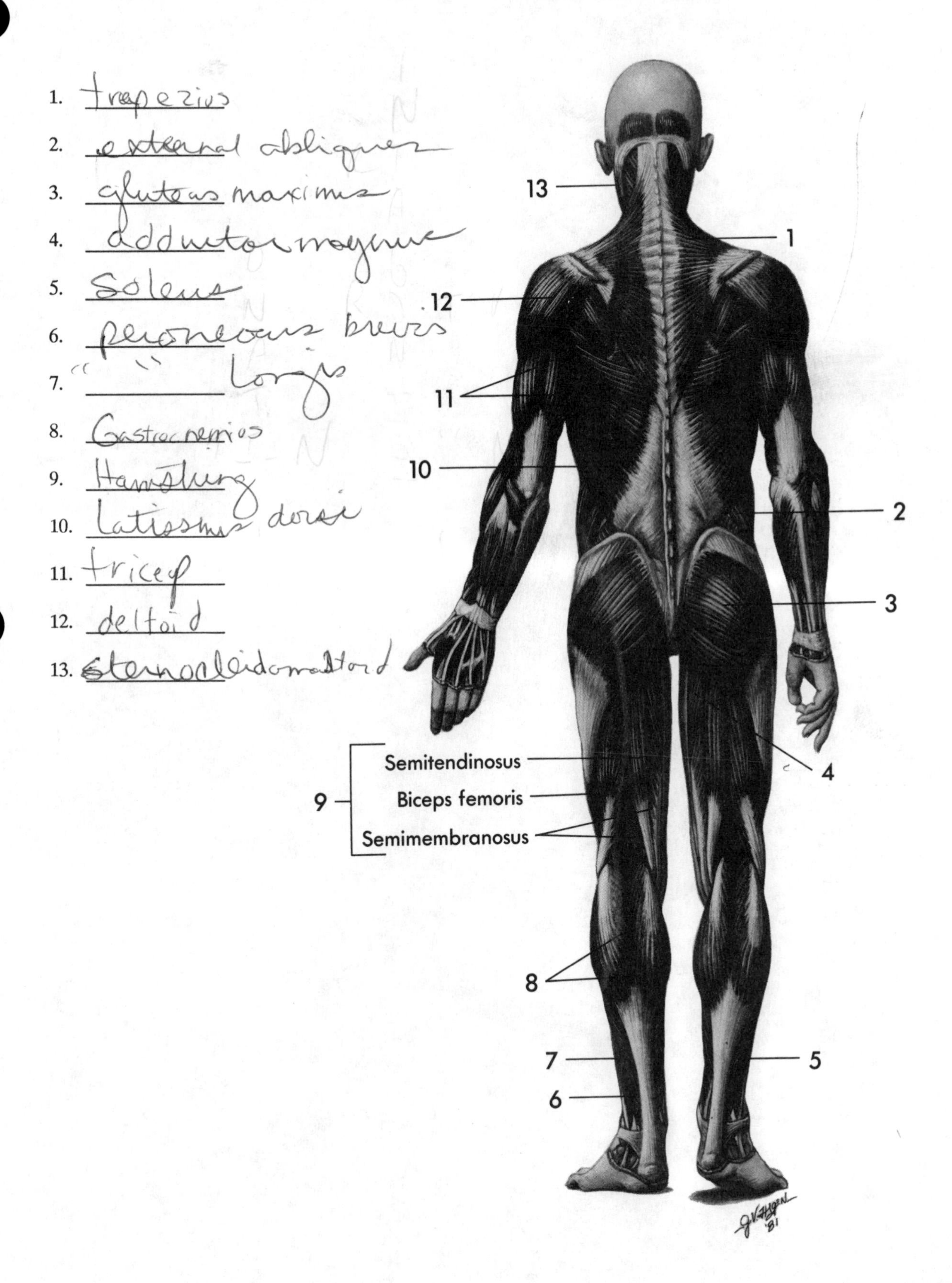

CHAPTER 7

The Nervous System

The nervous system organizes and coordinates the millions of impulses received each day to make communication with and enjoyment of our environment possible. The functioning unit of the nervous system is the neuron. Three types of neurons, sensory, motor, and interneurons, exist and are classified according to the direction in which they transmit impulses. Nerve impulses travel over routes made up of neurons and provide the rapid communication that is necessary for maintaining life. The central nervous system is made up of the spinal cord and brain. The spinal cord provides access to and from the brain by means of ascending and descending tracts. In addition, the spinal cord functions as the primary reflex center of the body. The brain can be subdivided for easier learning into the brain stem, cerebellum, diencephalon, and cerebrum. These areas provide the extraordinary network necessary to receive, interpret, and respond to the simplest or most complex impulses.

While you concentrate on this chapter, your body is performing a multitude of functions. Fortunately for us, the beating of the heart, the digestion of food, breathing, and most of our other day-to-day processes do not require our supervision or thought. They function automatically, and the division of the nervous system that regulates these functions is known as the autonomic nervous system.

The autonomic nervous system consists of two divisions called the sympathetic system and the parasympathetic system. The sympathetic system functions as an emergency system and prepares us for "fight" or "flight." The parasympathetic system dominates control of many visceral effectors under normal everyday conditions. Together, these two divisions regulate the body's automatic functions in an effort to assist with the maintenance of homeostasis. Your understanding of this chapter will alert you to the complexity and functions of the nervous system and the "automatic pilot" of your body—the autonomic system.

TOPICS FOR REVIEW

Before progressing to Chapter 8, you should review the organs and divisions of the nervous system, the structure and function of the major types of cells in this system, the structure and function of a reflex arc, and the transmission of nerve impulses. Your study should include the anatomy and physiology of the brain and spinal cord and the nerves that extend from these two areas.

Finally, an understanding of the autonomic nervous system and the specific functions of the subdivisions of this system are necessary to complete the review of this chapter.

ORGANS AND DIVISIONS OF THE NERVOUS SYSTEM
CELLS OF THE NERVOUS SYSTEM

Match the term on the left with the proper selection on the right.

Group A

__b__ 1. Sense organ
__c__ 2. Central nervous system
__d__ 3. Peripheral nervous system
__a__ 4. Autonomic nervous system

a. Subdivision of peripheral nervous system
b. Ear
c. Brain and spinal cord
d. Nerves that extend to the outlying parts of the body

Group B

__b__ 5. Dendrite
__d__ 6. Schwann cell
__c__ 7. Motor neuron
__a__ 8. Nodes of Ranvier
__f__ 9. Fascicles
__e__ 10. Epineurium

a. Indentations between adjacent Schwann cells
b. Branching projection of neuron
c. Also known as efferent
d. Forms myelin outside the CNS
e. Tough sheath that covers the whole nerve
f. Groups of wrapped axons

CELLS OF NERVOUS SYSTEM

Select the best choice for the following words and insert the correct letter in the answer blank.

(a) Neurons (b) Neuroglia

__A__ 11. Axon
__B__ 12. Connective tissue
__B__ 13. Astrocytes
__A__ 14. Sensory
__A__ 15. Conduct impulses
__B__ 16. Forms the myelin sheath around central nerve fibers
__B__ 17. Phagocytosis
__A__ 18. Efferent
__B__ 19. Multiple sclerosis
__A__ 20. Neurilemma

▸ If you have had difficulty with this section, review pages 132-136. ◂

REFLEX ARCS

Fill in the blanks.

21. The simplest kind of reflex arc is a __two neuron__ (patellar)

22. Three-neuron arcs consist of all three kinds of neurons, __motor__, __sensory__, and __inter neurons__

23. Impulse conduction in a reflex arc normally starts in __receptors__.

24. A ___synapse___ is the microscopic space that separates the axon of one neuron from the dendrites of another neuron.

25. A ___reflex___ is the response to impulse conduction over reflex arcs.

26. Contraction of a muscle that causes it to pull away from an irritating stimulus is known as the ___withdrawal reflex___

27. ___Saltory conduction___ provides a more rapid type of impulse travel than is possible in nonmyelinated sections.

28. All ___interneurons___ lie entirely within the gray matter of the central nervous system.

29. In a patellar reflex, the nerve impulses that reach the quadriceps muscle (the effector) result in the classic ___knee-jerk___ response.

30. ___Gray matter___ forms the H-shaped inner core of the spinal cord.

▸ If you have had difficulty with this section, review pages 137-138. ◂

NERVE IMPULSES

Circle the correct answer.

31. Nerve impulses (do or *do not* [circled]) continually race along every nerve cell's surface.

32. When a stimulus acts on a neuron, it (increases or decreases) the permeability of the stimulated point of its membrane to sodium ions.

33. An inward movement of positive ions leaves a/an (lack or excess) of negative ions outside.

34. The plasma membrane of the (presynaptic neuron or postsynaptic neuron) makes up a portion of the synapse.

35. A synaptic knob is a tiny bulge at the end of the (presynaptic or postsynaptic) neuron's axon.

36. Acetylcholine is an example of a (neurotransmitter or protein molecule receptor).

37. Neurotransmitters are chemicals that allow neurons to (communicate or reproduce) with one another.

38. Neurotransmitters are distributed (randomly or specifically) into groups of neurons.

39. Catecholamines may play a role in (sleep or reproduction).

40. Endorphins and enkephalins are neurotransmitters that inhibit conduction of (fear or pain) impulses.

▸ If you have had difficulty with this section, review pages 138-141. ◂

DIVISIONS OF THE BRAIN

Circle the correct choice.

41. The portion of the brain stem that joins the spinal cord to the brain is the:

    a. Pons
    b. Cerebellum
    c. Diencephalon
    d. Hypothalamus
    e. Medulla

42. Which one of the following is *not* a function of the brain stem?

    a. Conducts sensory impulses from the spinal cord to the higher centers of the brain.
    b. Conducts motor impulses from the cerebrum to the spinal cord.
    c. Controls heartbeat, respiration, and blood vessel diameter.
    d. Contains centers for speech and memory.

43. Which one of the following is *not* part of the diencephalon?

    a. Cerebrum
    b. Thalamus
    c. Pituitary gland
    d. Third ventricle gray matter

44. ADH is produced by the:

    a. Pituitary gland
    b. Medulla
    c. Mammillary bodies
    d. Third ventricle
    e. Hypothalamus

45. Which one of the following is *not* a function of the hypothalamus?

    a. It controls the rate of heartbeat.
    b. It controls the constriction and dilation of blood vessels.
    c. It controls the contraction of the stomach and intestines.
    d. It produces releasing hormones that affect the posterior pituitary.
    e. All of the above are functions of the hypothalamus.

46. Which one of the following parts of the brain helps in the association of sensations with emotions, as well as aiding in the arousal or alerting mechanism?

    a. Pons
    b. Hypothalamus
    c. Cerebellum
    d. Thalamus
    e. None of the above is correct

47. Which of the following is *not* true of the cerebrum?

    a. Its lobes correspond to the bones that lie over them.
    b. Its grooves are called gyri.
    c. Most of its gray matter lies on the surface of the cerebrum.
    d. Its outer region is called the cerebral cortex.
    e. Its two hemispheres are connected by a structure called the corpus callosum.

48. Which one of the following is *not* a function of the cerebrum?

a. Willed movement
b. Consciousness
c. Memory
d. Conscious awareness of sensations
e. All of the above are functions of the cerebrum

49. The area of the cerebrum responsible for the perception of sound lies in the ________________ lobe.

a. Frontal
b. Temporal
c. Occipital
d. Parietal

50. Visual perception is located in the ______________________ lobe.

a. Frontal
b. Temporal
c. Parietal
d. Occipital
e. None of the above is correct

51. Which one of the following is *not* a function of the cerebellum?

a. Maintains equilibrium
b. Helps produce smooth, coordinated movements
c. Helps maintain normal postures
d. Associates sensations with emotions

52. Within the interior of the cerebrum are a few islands of gray matter known as:

a. Fissures
b. Basal ganglia
c. Gyri
d. Myelin

53. A cerebrovascular accident is commonly referred to as (a)

a. Stroke
b. Parkinson's disease
c. Tumor
d. Multiple sclerosis

54. Parkinson's disease is a disease of the:

a. Myelin
b. Axons
c. Neuroglia
d. Basal ganglia

55. The largest section of the brain is the

a. Cerebellum
b. Pons
c. Cerebrum
d. Midbrain

▸ If you have had difficulty with this section, review pages 142-146. ◂

SPINAL CORD

If the statement is true, write **T** *on the answer blank. If the statement is false, correct the statement by circling the incorrect term and inserting the correct term in the answer blank.*

_________56. The spinal cord is approximately 24 to 25 inches long.
_________57. The spinal cord ends at the bottom of the sacrum.
_________58. The extension of the meninges beyond the cord is convenient for performing CAT scans without danger of injuring the spinal cord.
_________59. Bundles of myelinated nerve fibers—dendrites—make up the white outer columns of the spinal cord.
_________60. Ascending tracts conduct impulses up the cord to the brain and descending tracts conduct impulses down the cord from the brain.
_________61. Tracts are functional organizations in that all the axons that compose a tract serve several functions.
_________62. A loss of sensation caused by a spinal cord injury is called paralysis.

▸ If you have had difficulty with this section, review pages 146-149. ◂

COVERINGS AND FLUID SPACES OF BRAIN AND SPINAL CORD

Circle the one that does not belong.

63. Meninges	Pia mater	Ventricles	Dura mater
64. Arachnoid	Middle layer	CSF	Cobweblike
65. CSF	Ventricles	Subarachnoid space	Pia mater
66. Tough	Vertebral canal	Dura mater	Choroid plexus
67. Brain tumor	Subarachnoid space	CSF	Fourth lumbar vertebra

CRANIAL NERVES

68. Fill in the missing areas on the chart below.

NERVE		CONDUCT IMPULSES	FUNCTION
I		From nose to brain	Sense of smell
II	Optic	From eye to brain	
III	Oculomotor		Eye movements
IV		From brain to external eye muscles	Eye movements
V	Trigeminal	From skin and mucous membrane of head and from teeth to brain; also from brain to chewing muscles	
VI	Abducens		Turning eyes outward
VII	Facial	From taste buds of tongue to brain; from brain to face muscles	
VIII		From ear to brain	Hearing; sense of balance
IX	Glossopharyngeal		Sensations of throat, taste, swallowing movements, secretion of saliva
X		From throat, larynx, and organs in thoracic and abdominal cavities to brain; also from brain to muscles of throat and to organs in thoracic and abdominal cavities	Sensations of throat, larynx, and of thoracic and abdominal organs; swallowing, voice production, slowing of heartbeat, acceleration of peristalsis (gut movements)
XI	Spinal accessory	From brain to certain shoulder and neck muscles	
XII		From brain to muscles of tongue	Tongue movements

▸ If you have had difficulty with this section, review page 153, Table 7-2. ◂

CRANIAL NERVES
SPINAL NERVES

Select the best choice for the following words and insert the correct letter in the answer blank.

(a) Cranial nerves (b) Spinal nerves

_____69. 12 pairs
_____70. Dermatome
_____71. Vagus
_____72. Shingles
_____73. 31 pairs
_____74. Optic
_____75. C1
_____76. Plexus

▸ If you have had difficulty with this section, review pages 152-156. ◂

DEFINITIONS
NAMES OF DIVISIONS

Match the term on the left with the proper selection on the right.

_____77. Autonomic nervous system
_____78. Autonomic neurons
_____79. Preganglionic neurons
_____80. Visceral effectors
_____81. Sympathetic system
_____82. Somatic nervous system

a. Divisions of ANS
b. Tissues to which autonomic neurons conduct impulses
c. Voluntary actions
d. Regulates body's involuntary functions
e. Motor neurons that make up the ANS
f. Conduct impulses between the spinal cord and a ganglion

SYMPATHETIC NERVOUS SYSTEM
PARASYMPATHETIC NERVOUS SYSTEM

Circle the correct choice.

83. Dendrites and cell bodies of sympathetic preganglionic neurons are located in the:

 a. Brain stem and sacral portion of the spinal cord
 b. Sympathetic ganglia
 c. Gray matter of the thoracic and upper lumbar segments of the spinal cord
 d. Ganglia close to effectors

84. Which of the following is *not* correct?

 a. Sympathetic preganglionic neurons have their cell bodies located in the lateral gray column of certain parts of the spinal cord.
 b. Sympathetic preganglionic axons pass along the dorsal root of certain spinal nerves.
 c. There are synapses within sympathetic ganglia.
 d. Sympathetic responses are usually widespread, involving many organs.

85. Another name for the parasympathetic nervous system is:

a. Thoracolumbar
b. Craniosacral
c. Visceral
d. ANS
e. Cholinergic

86. Which statement is *not* correct?

a. Sympathetic postganglionic neurons have their dendrites and cell bodies in sympathetic ganglia or collateral ganglia.
b. Sympathetic ganglions are located in front of and at each side of the spinal column.
c. Separate autonomic nerves distribute many sympathetic postganglionic axons to various internal organs.
d. Very few sympathetic preganglionic axons synapse with postganglionic neurons.

87. Sympathetic stimulation usually results in:

a. Response by numerous organs
b. Response by only one organ
c. Increased peristalsis
d. Constriction of pupils

88. Parasympathetic stimulation frequently results in:

a. Response by only one organ
b. Responses by numerous organs
c. The fight or flight syndrome
d. Increased heartbeat

Choose the correct response and insert the letter in answer blanks.

(a) Sympathetic control (b) Parasympathetic control

_____ 89. Constricts pupils
_____ 90. "Goose pimples"
_____ 91. Increases sweat secretion
_____ 92. Increases secretion of digestive juices
_____ 93. Constricts blood vessels
_____ 94. Slows heartbeat
_____ 95. Relaxes bladder
_____ 96. Increases epinephrine secretion
_____ 97. Increases peristalsis
_____ 98. Stimulates lens for near vision

▸ If you have had difficulty with this section, review pages 156-160. ◂

AUTONOMIC NEUROTRANSMITTERS
AUTONOMIC NERVOUS SYSTEM AS A WHOLE

Fill in the blanks.

99. Sympathetic preganglionic axons release the neurotransmitter ____________________.

100. Axons that release norepinephrine are classified as ______________.

101. Axons that release acetylcholine are classified as ______________.

102. The function of the autonomic nervous system is to regulate the body's involuntary functions in ways that maintain or restore __________.

103. Your __________________ is determined by the combined forces of the sympathetic and parasympathetic nervous system.

104. According to some physiologists, meditation leads to ________________ sympathetic activity and changes opposite to those of the fight or flight syndrome.

▸ If you have had difficulty with this section, review pages 160-161. ◂

Unscramble the words.

105. RONNESU

106. APSYENS

107. CIATUNOMO

108. SHTOMO ULMSEC

Take the circled letters, unscramble them, and fill in the statement.

What the man hoped the IRS agent would be during his audit.

109.

APPLYING WHAT YOU KNOW

110. Mr. Hemstreet suffered a cerebrovascular accident and it was determined that the damage affected the left side of his cerebrum. On which side of his body will he most likely notice any paralysis?

111. Baby Hansen was born with an excessive accumulation of cerebrospinal fluid in the ventricles. A catheter was placed in the ventricle and the fluid was drained by means of a shunt into the circulatory bloodstream. What condition does this medical history describe?

112. Mrs. Gordon looked out her window to see a man trapped under the wheel of a car. Although slightly built, Mrs. Gordon rushed to the car, lifted it, and saved the man underneath the wheel. What division of the autonomic nervous system made this seemingly impossible task possible?

113. Cassidy's heart raced and her palms became clammy as she watched the monster at the local theater. When the movie was over, however, she told her friends that she was not afraid at all. She appeared to be as calm as before the movie. What division of the autonomic nervous system made this possible?

114. Bill was going to his boss for his annual evaluation. He is planning to ask for a raise and hopes the evaluation will be good. Which subdivision of the autonomic nervous system will be active during this conference? Should he have a large meal before his appointment? Support your answer with facts from the chapter.

DID YOU KNOW?

Although all pain is felt and interpreted in the brain, it has no pain sensation itself — even when cut!

THE NERVOUS SYSTEM

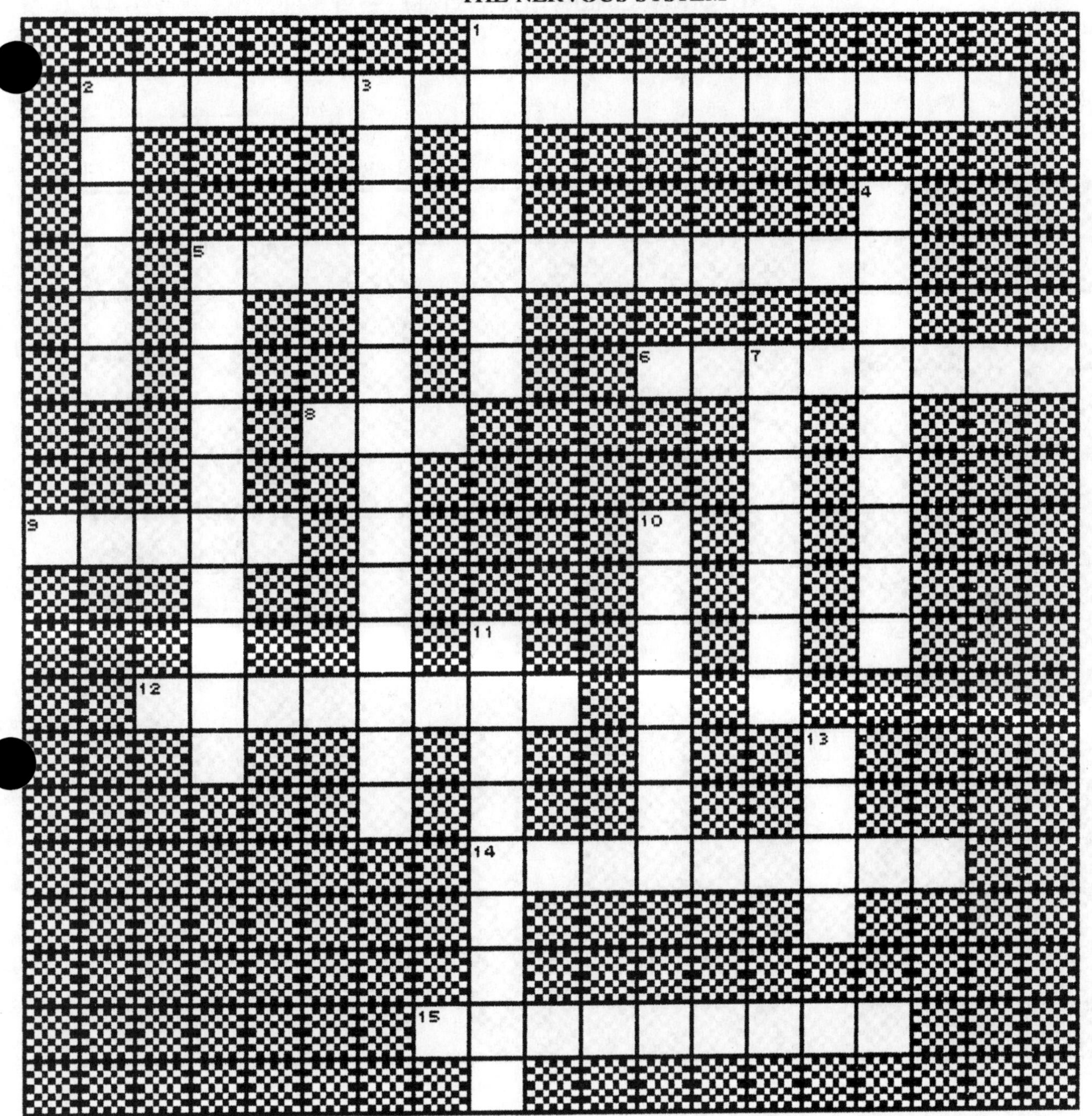

ACROSS

2. Myelin disorder
5. Neurotransmitter
6. Cluster of nerve cell bodies outside the central nervous system
8. Peripheral nervous system (abbrev.)
9. Bundle of axons located within the CNS
12. Pia mater
14. Peripheral beginning of a sensory neuron's dendrite
15. Astrocytes

DOWN

1. Area of the brain stem
2. Fatty substance found around some nerve fibers
3. Neurons that conduct impulses from a ganglion
4. Two neuron arc
5. Neuroglia
7. Nerve cells
10. Where impulses are transmitted from one neuron to another
11. Transmits impulses toward the cell body
13. Transmits impulses away from the cell body

AUTONOMIC NERVOUS SYSTEM

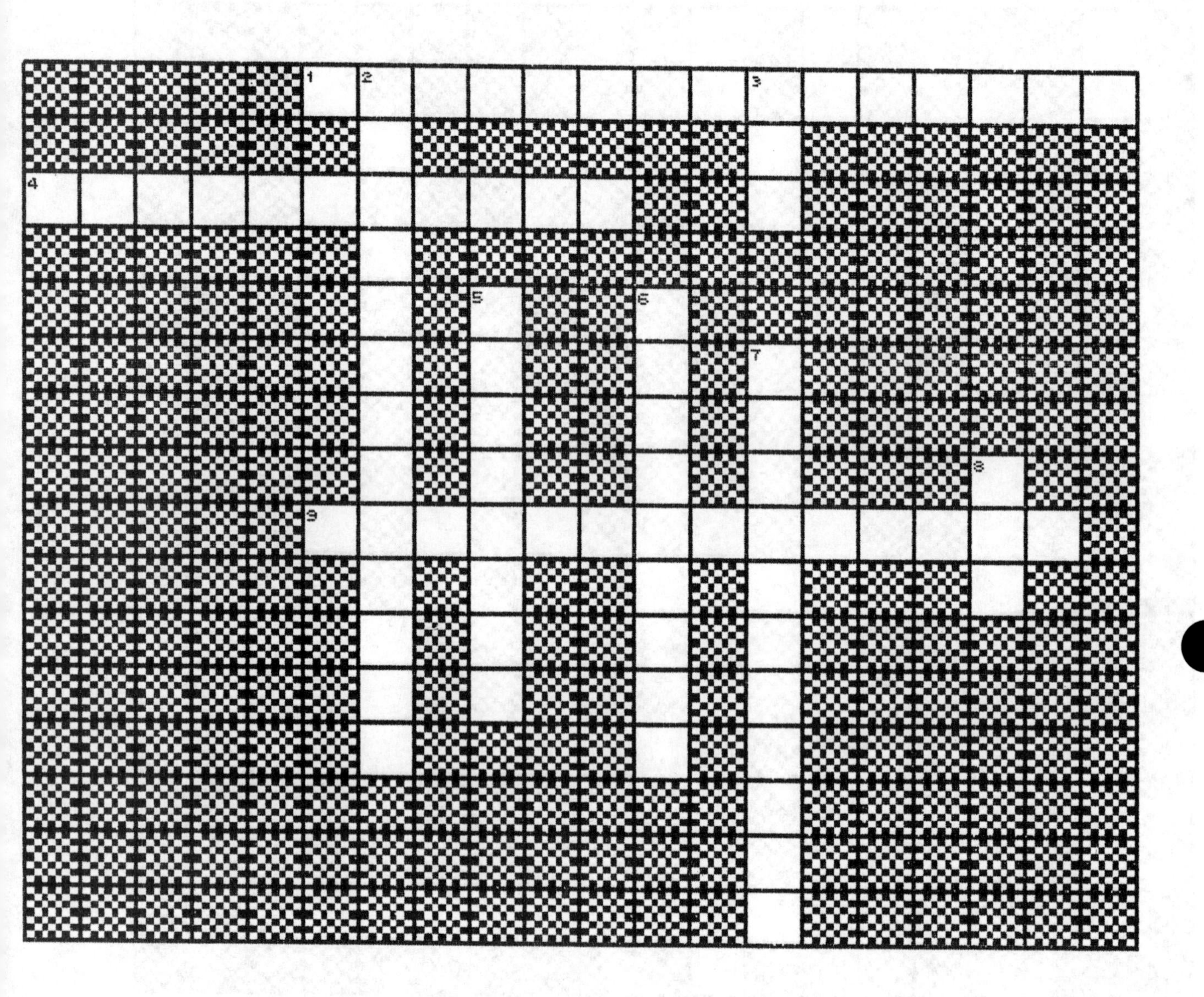

ACROSS

1. System that controls visceral effectors under normal conditions
4. Type of fibers which release acetylcholine
9. Adrenergic fibers release this neurotransmitter

DOWN

2. Neurotransmitter released by cholinergic fibers
3. Autonomic nervous system (abbrev.)
5. Type of effect which is autonomic
6. Also known as visceral effectors
7. System that functions as an emergency one
8. Parasympathetic nervous system (abbrev.)

NEURON

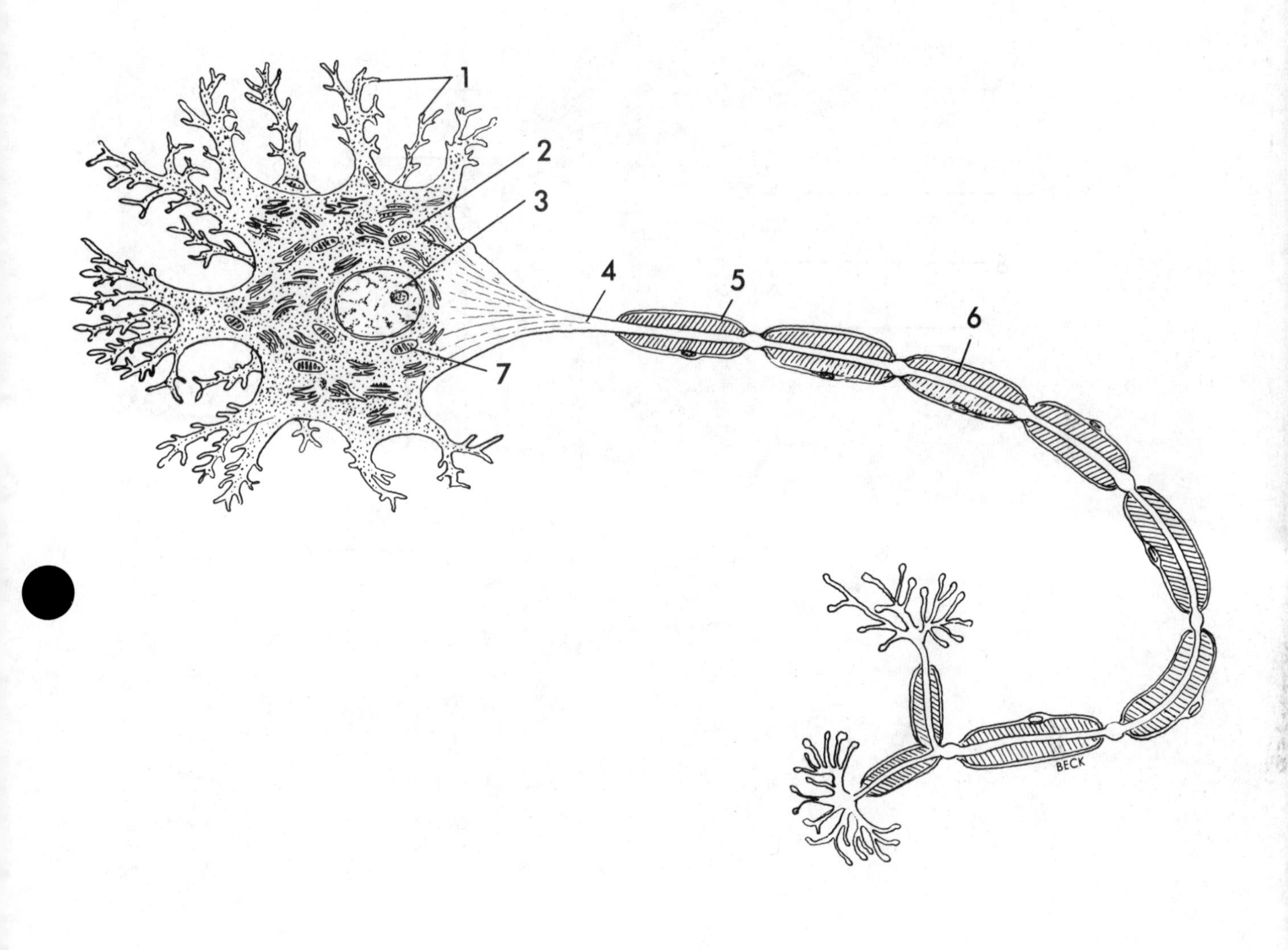

1. ______________________________
2. ______________________________
3. ______________________________
4. ______________________________
5. ______________________________
6. ______________________________
7. ______________________________

CROSS-SECTION OF SPINAL CORD

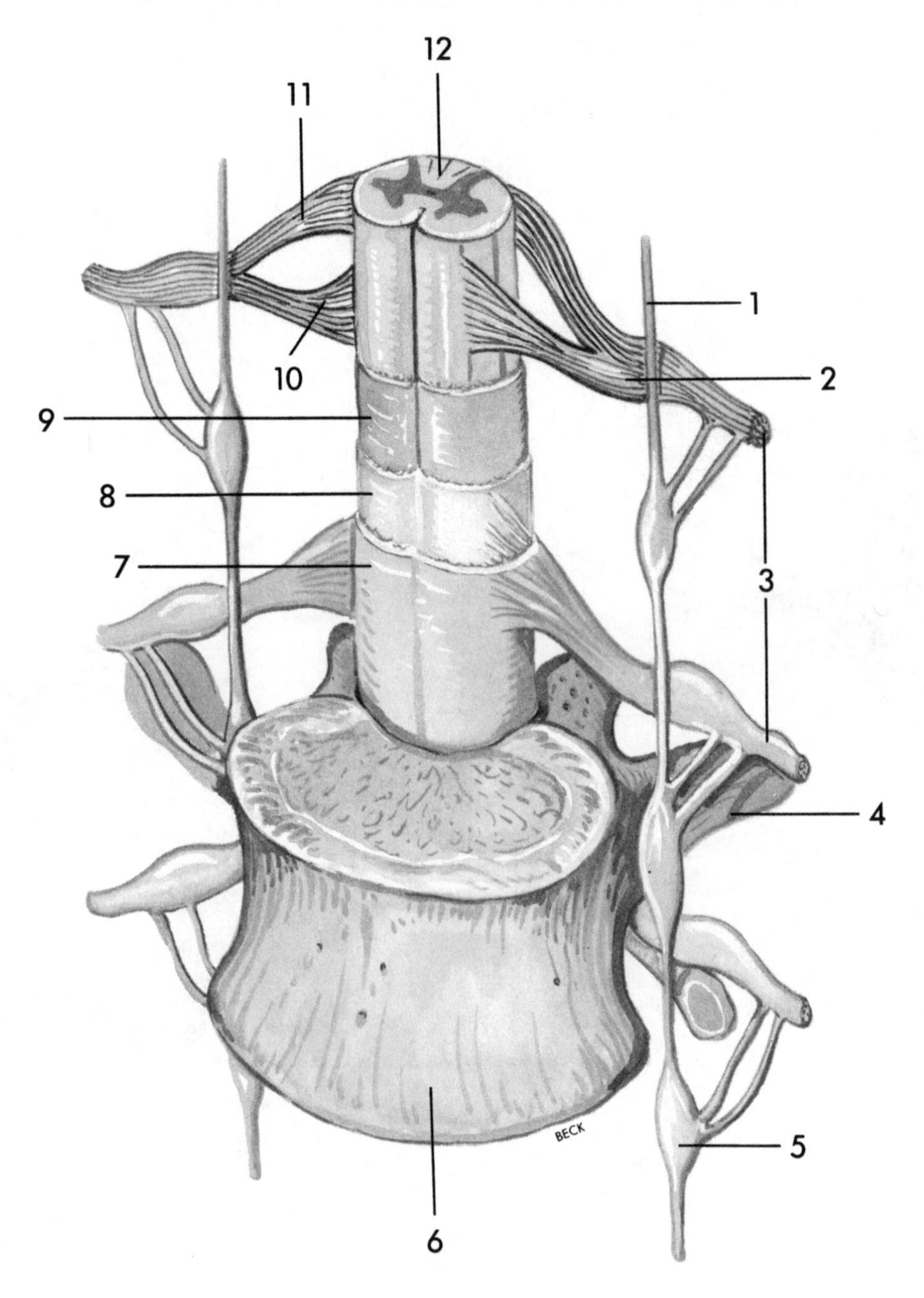

1. ______________________
2. ______________________
3. ______________________
4. ______________________
5. ______________________
6. ______________________
7. ______________________
8. ______________________
9. ______________________
10. ______________________
11. ______________________
12. ______________________

NEURAL PATHWAY INVOLVED IN THE PATELLAR REFLEX

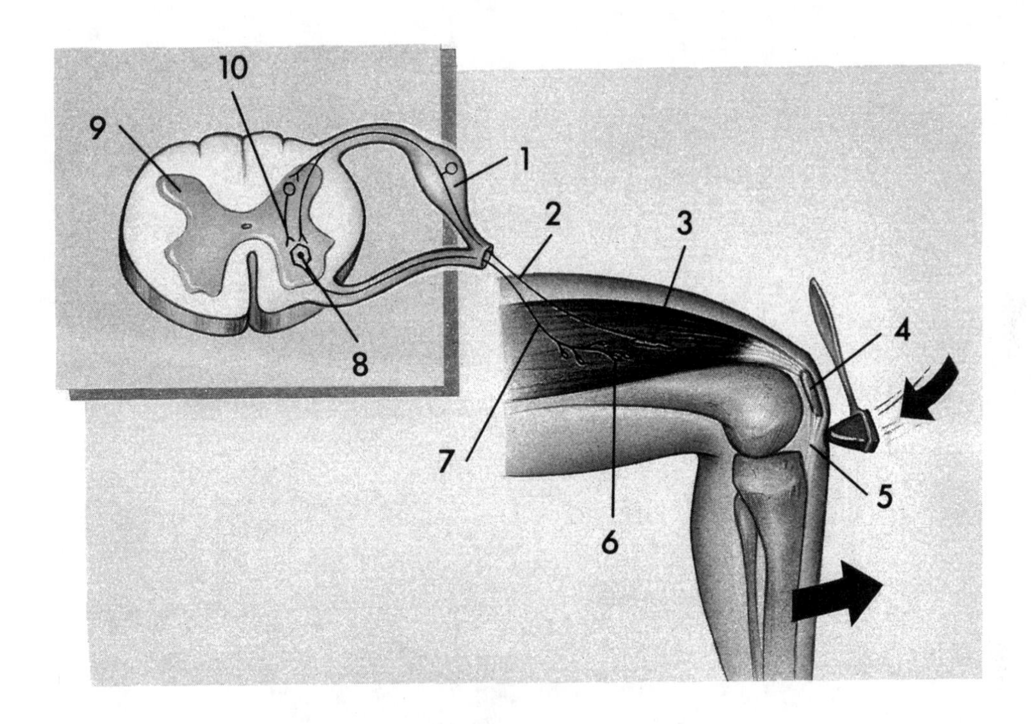

1. ______________________
2. ______________________
3. ______________________
4. ______________________
5. ______________________
6. ______________________
7. ______________________
8. ______________________
9. ______________________
10. ______________________

THE CEREBRUM

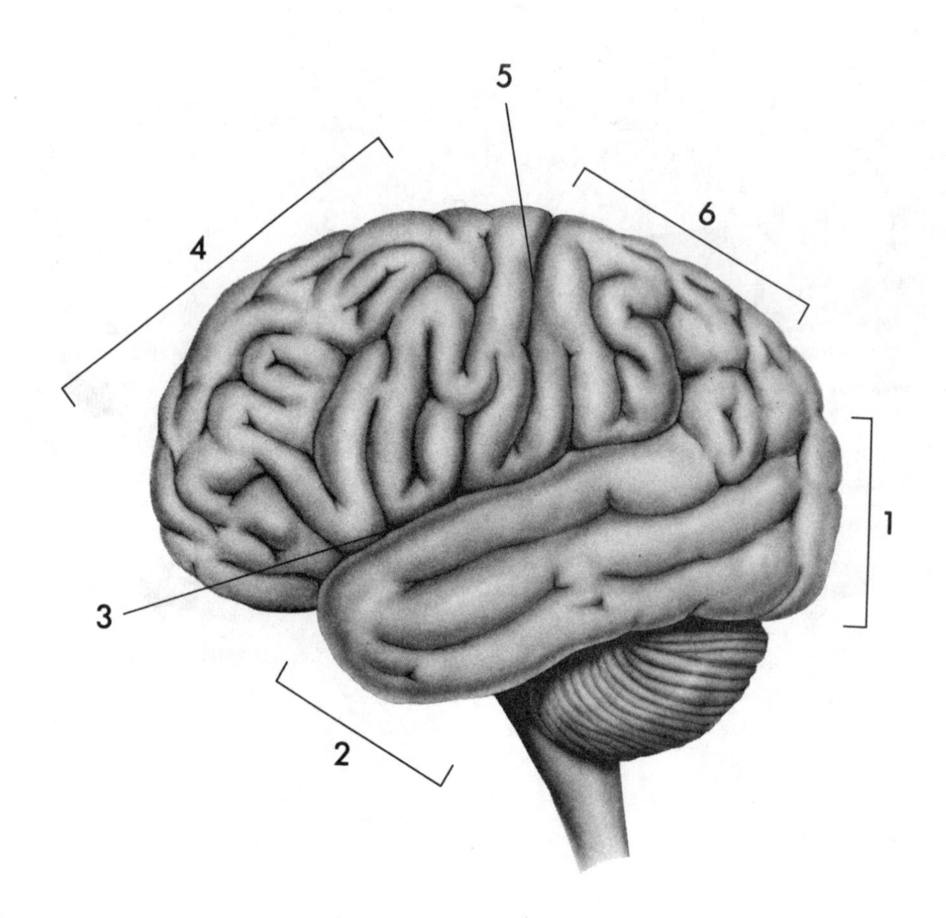

1. ______________________

2. ______________________

3. ______________________

4. ______________________

5. ______________________

6. ______________________

SAGITTAL SECTION OF THE CENTRAL NERVOUS SYSTEM

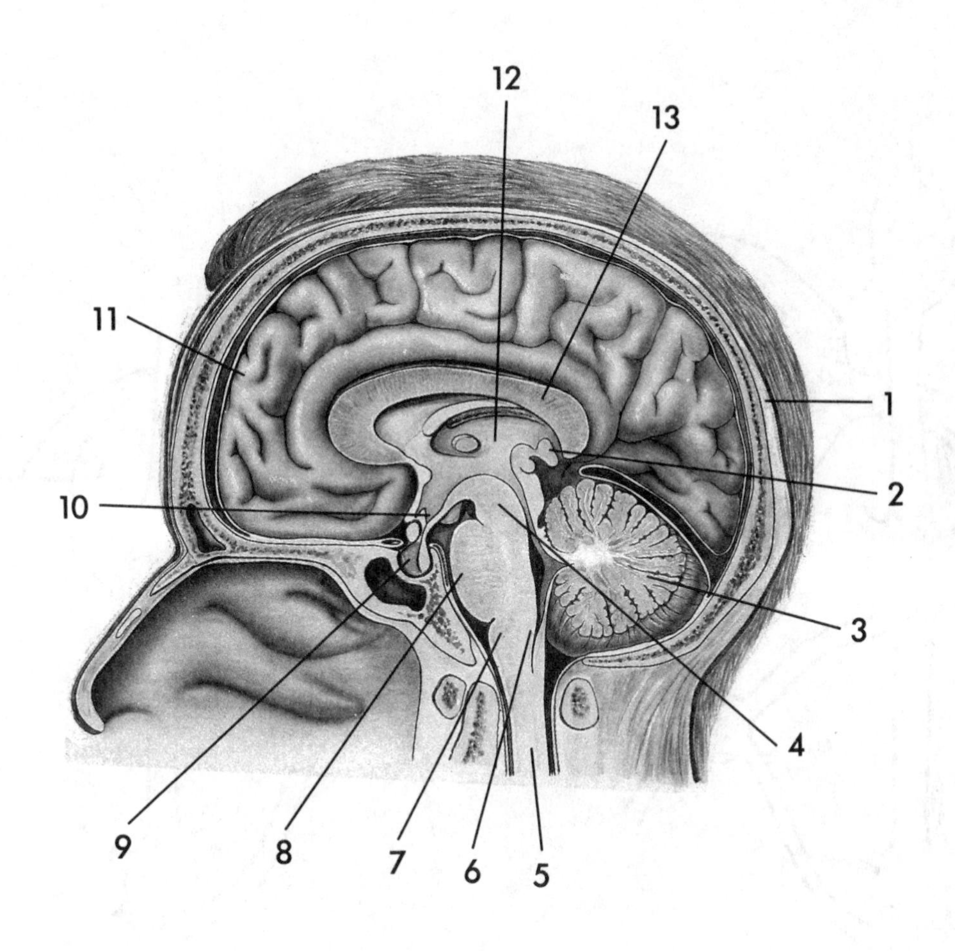

1. ______________________________
2. ______________________________
3. ______________________________
4. ______________________________
5. ______________________________
6. ______________________________

6. ______________________________
7. ______________________________
8. ______________________________
9. ______________________________
10. ______________________________
11. ______________________________

NEURON PATHWAYS

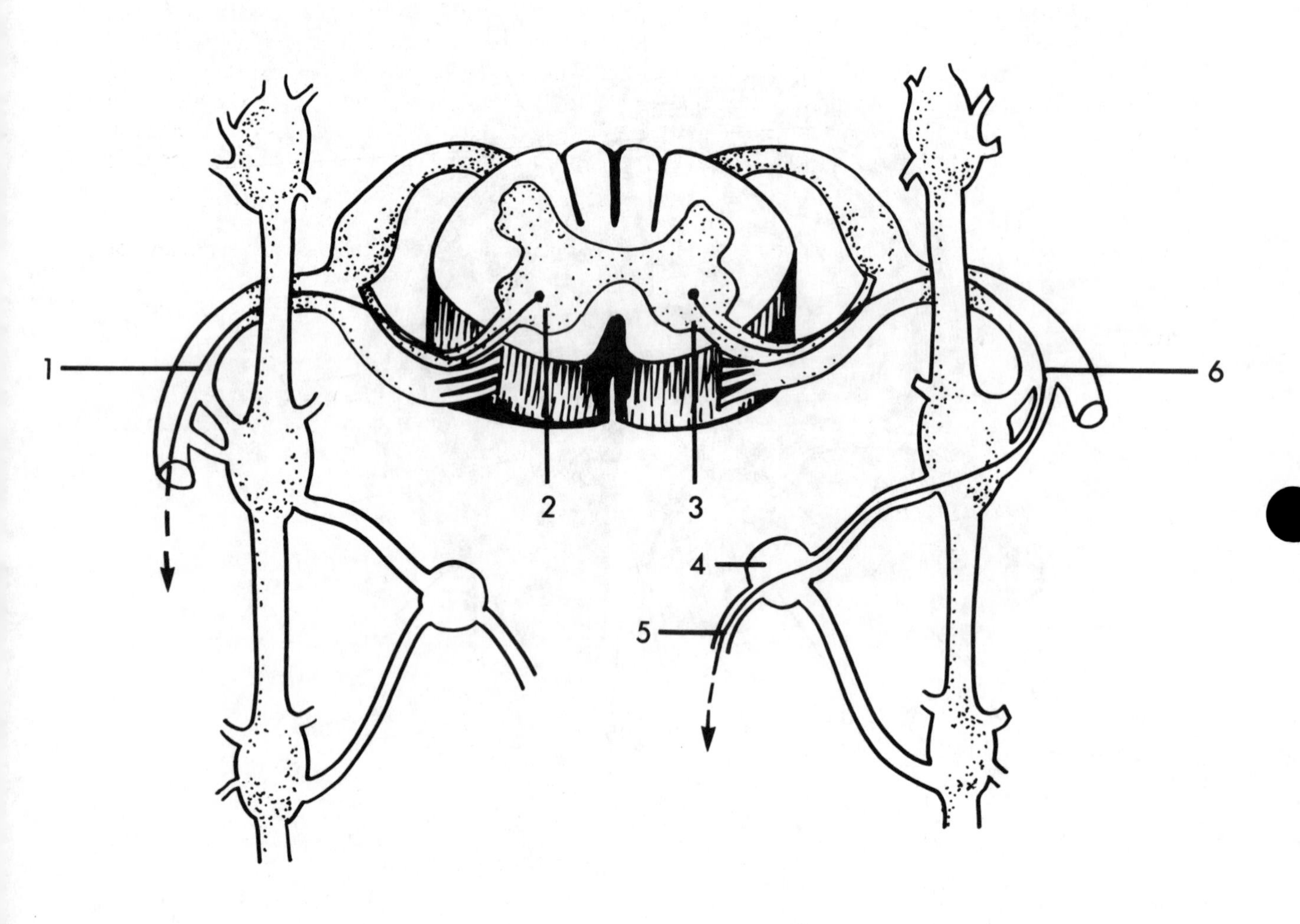

1. ______________________
2. ______________________
3. ______________________
4. ______________________
5. ______________________
6. ______________________

CHAPTER 8

The Senses

Consider this scene for a moment. You are walking along a beautiful beach watching the sunset. You notice the various hues and are amazed at the multitude of shades that cover the sky. The waves are indeed melodious as they splash along the shore and you wiggle your feet with delight as you sense the warm, soft sand trickling between your toes. You sip on a soda and then inhale the fresh salt air as you continue your stroll along the shore. It is a memorable scene, but one that would not be possible without the assistance of your sense organs. The sense organs pick up messages that are sent over nerve pathways to specialized areas in the brain for interpretation. They make communication with and enjoyment of the environment possible. The visual, auditory, tactile, olfactory, and gustatory sense organs not only protect us from danger but also add an important dimension to our daily pleasures of life.

Your study of this chapter will give you an understanding of another one of the systems necessary for homeostasis and survival.

TOPICS FOR REVIEW

Before progressing to Chapter 9, you should review the classification of sense organs and the process for converting a stimulus into a sensation. Your study should also include an understanding of the special sense organs and the general sense organs.

CLASSIFICATION OF SENSE ORGANS
CONVERTING A STIMULUS INTO A SENSATION

Match the term on the left with the proper selection on the right.

____ 1. Special sense organ
____ 2. General sense organ
____ 3. Nose
____ 4. Krause's end-bulbs
____ 5. Taste buds

a. Olfactory cells
b. Meissner's corpuscles
c. Chemoreceptor
d. Eye
e. Touch

▸ If you have had difficulty with this section, review pages 169-178. ◂

SPECIAL SENSE ORGANS

Eye

Circle the correct choice.

6. The "white" of the eye is more commonly called the:

a. Choroid
b. Cornea
c. Sclera
d. Retina
e. None of the above is correct

7. The "colored" part of the eye is known as the:

a. Retina
b. Cornea
c. Pupil
d. Sclera
e. Iris

8. The transparent portion of the sclera, referred to as the "window" of the eye, is the:

a. Retina
b. Cornea
c. Pupil
d. Iris

9. The mucous membrane that covers the front of the eye is called the:

a. Cornea
b. Choroid
c. Conjunctiva
d. Ciliary body
e. None of the above is correct

10. The structure that can contract or dilate to allow more or less light to enter the eye is the:

a. Lens
b. Choroid
c. Retina
d. Cornea
e. Iris

11. When the eye is looking at objects far in the distance, the lens is Slighty Curve and the ciliary muscle is relaxed.

a. Rounded, contracted
b. Rounded, relaxed
c. Slightly rounded, contracted
d. Slightly curved, relaxed
e. None of the above is correct

12. The lens of the eye is held in place by the:

a. Ciliary muscle
b. Aqueous humor
c. Vitreous humor
d. Cornea

13. When the lens loses its elasticity and can no longer bring near objects into focus, the condition is known as:

a. Glaucoma
b. Presbyopia
c. Astigmatism
d. Strabismus

14. The fluid in front of the lens that is constantly being formed, drained, and replaced in the anterior cavity is the:

a. Vitreous humor
b. Protoplasm
c. Aqueous humor
d. Conjunctiva

15. If drainage of the aqueous humor is blocked, the internal pressure within the eye will increase and a condition known as ________________ could occur.

a. Presbyopia
b. Glaucoma
c. Color blindness
d. Cataracts

16. The rods and cones are the visual receptors and are located on the:

a. Sclera
b. Cornea
c. Choroid
d. Retina

17. Photoreception is the sense of:

a. Color
b. Smell
c. Position and movement
d. Distance

18. If our eyes are abnormally elongated, the image focuses in front of the retina and a condition known as ________________ occurs.

a. Hyperopia
b. Cataracts
c. Night blindness
d. Myopia

▸ If you have had difficulty with this section, review pages 168-172. ◂

Ear

Select the best answer from the choices given and insert the letter in the answer blank.

(a) External ear (b) Middle ear (c) Inner ear

b 19. Malleus
c 20. Perilymph
b 21. Incus
a 22. Ceruminous glands
c 23. Cochlea
a 24. Auditory canal
c 25. Semicircular canals
b 26. Stapes
a 27. Tympanic membrane
c 28. Organ of Corti

Fill in the blanks.

29. The external ear has two parts: the auricle (pinna) and the external auditory canal

30. Another name for the tympanic membrane is the eardrum.

31. The bones of the middle ear are referred to, collectively, as ossicles.

32. The stapes presses against a membrane that covers a small opening, the oval window.

33. A middle ear infection is called otitis media.

34. The ____________________ is located adjacent to the oval window between the semicircular canals and the cochlea.

35. Located within the semicircular canals and the vestibule are _____________ for balance and equilibrium.

36. The sensory cells in the ________________ ___________________ are stimulated when movement of the head causes the endolymph to move.

▸ If you have had difficulty with this section, review pages 173-175. ◂

TASTE RECEPTORS
SMELL RECEPTORS
GENERAL SENSE ORGANS

Circle the correct answer.

37. Structures known as (papillae or olfactory cells) are found on the tongue.

38. Nerve impulses generated by stimulation of taste buds travel primarily through two (cranial or spinal) nerves.

39. To be detected by olfactory receptors chemicals must be dissolved in the watery (mucus or plasma) that lines the nasal cavity.

40. The pathways taken by olfactory nerve impulses and the areas where these impulses are interpreted are closely associated with areas of the brain important in (hearing or memory).

41. A sense of position and movement is known as (proprioception or mechanoreception).

▸ If you have had difficulty with this section, review pages 176-178. ◂

APPLYING WHAT YOU KNOW

42. Mr. Nay was an avid swimmer and competed regularly in his age group. He had to withdraw from the last competition due to an infection of his ear. Antibiotics and analgesics were prescribed by the doctor. What is the medical term for his condition?

43. Mrs. Metheny loved the out-of-doors and spent a great deal of her spare time basking in the sun on the beach. Her physician suggested that she begin wearing sunglasses regularly when he noticed milky spots beginning to appear on Mrs. Metheny's lenses. What condition was Mrs. Metheny's physician trying to prevent from occurring?

44. Amanda repeatedly became ill with throat infections during her first few years of school. Lately, however, she has noticed that whenever she has a throat infection, her ears become very sore also. What might be the cause of this additional problem?

45. Jeremy was hit in the nose with a baseball during practice. His sense of smell was temporarily gone. What nerve receptors were damaged during the injury?

46. WORD FIND

Can you find the terms from this chapter listed below in the box of letters? Words may be spelled top to bottom, bottom to top, right to left, left to right, or diagonally.

```
R C O Y P J A V I T C N U J N O C M X P
B R R O S S U C N I R D A O F X M Y I L
O L F A C T O R Y A R X S M F N X H V H
I R Y G F I Y T N K H B L W P A S E S J
V C A T A R A C T S X P K H P I Z N L A
G U S T A T O R Y B I Z O Y M H T T Y Q
Y Y H P A P I L L A E T A R D C X N J S
U Y T D L N C E L E O I E Q P A S W L O
V O C O C H L E A P P C E S E T X W D C
J C E R U M E N I O E U X E J S E A J C
Y G A E F Y X G Y P R D G N W U N V L F
X C O N E S M B T X R B X S A E J V Q E
H T M Y D E S O A I P O R E P Y H W E Z
Y W V N N E R J D P E A A S R F V A F S
S I Z T R S W H W C V Z I Q X H H K J O
I I R P M E C H A N O R E C E P T O R S
N O I T C A R F E R R N G Q Q E Y E W A
K S A W J A M T M X E W M S Q V F D R T
P J U G L G R O D S V C U G A A G N F R
E S Z I W Y T L G Q I M C A N F F C Q M
```

LIST OF WORDS

SENSES
CONJUNCTIVA
CONES
REFRACTION
CERUMEN
EUSTACHIAN
OLFACTORY
RECEPTORS
PRESBYOPIA
PHOTOPIGMENT
MECHANORECEPTORS
INCUS
GUSTATORY
EYE
RODS
CATARACTS
HYPEROPIA
COCHLEA
PAPILLAE

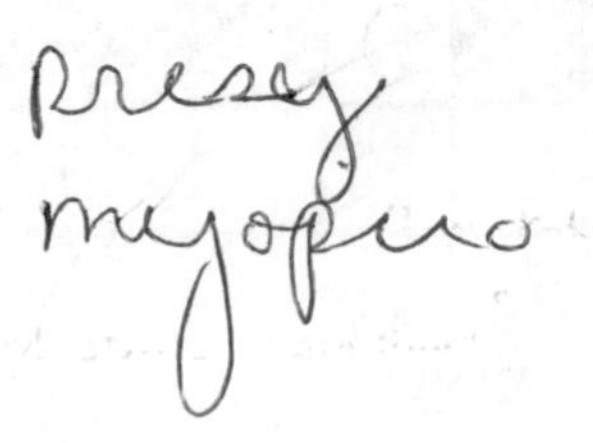

THE SENSES

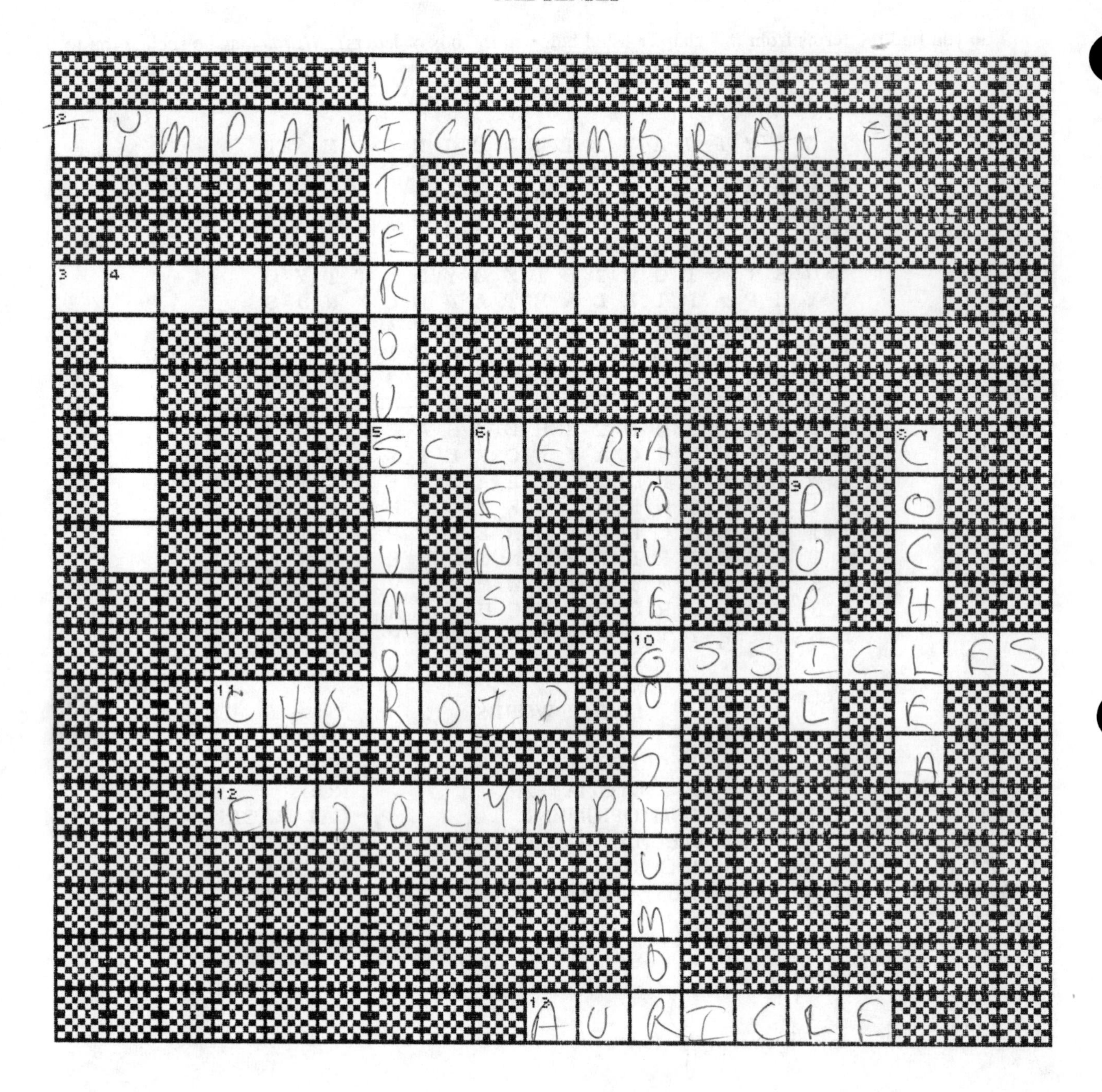

ACROSS

2. Eardrum (2 words)
3. Receptors for balance in semicircular canals (2 words)
5. White of the eye
10. Bones of the middle ear
11. Front part of this coat is the ciliary muscle and iris
12. Membranous labyrinth filled with this fluid
13. External ear

DOWN

1. Located in posterior cavity
4. Innermost layer of eye
6. Transparent body behind pupil
7. Located in anterior cavity in front of lens (2 words)
8. Organ of Corti located here
9. Hole in center of iris

EYE

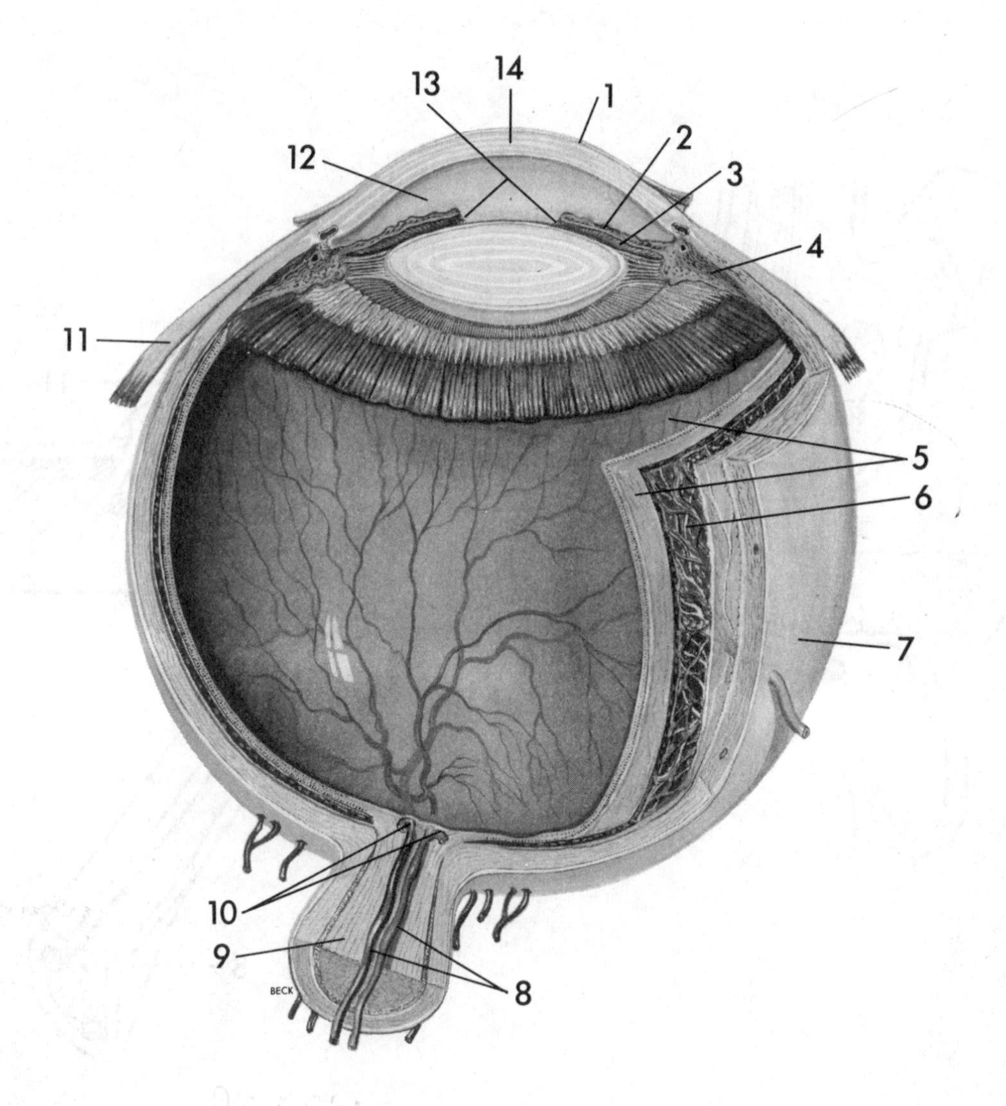

1. Conjuctiva
2. iris
3. ~~lens ciliary muscle~~ posterior cavity
4. ~~conjuctiva~~ ciliary muscle
5. Retina
6. choroid layer
7. Sclera
8. optic nerve
9. central retina artery + vein
10. optic disc (blind spot)
11. medial rectus muscle
12. anterior cavity
13. pupil
14. cornea

EAR

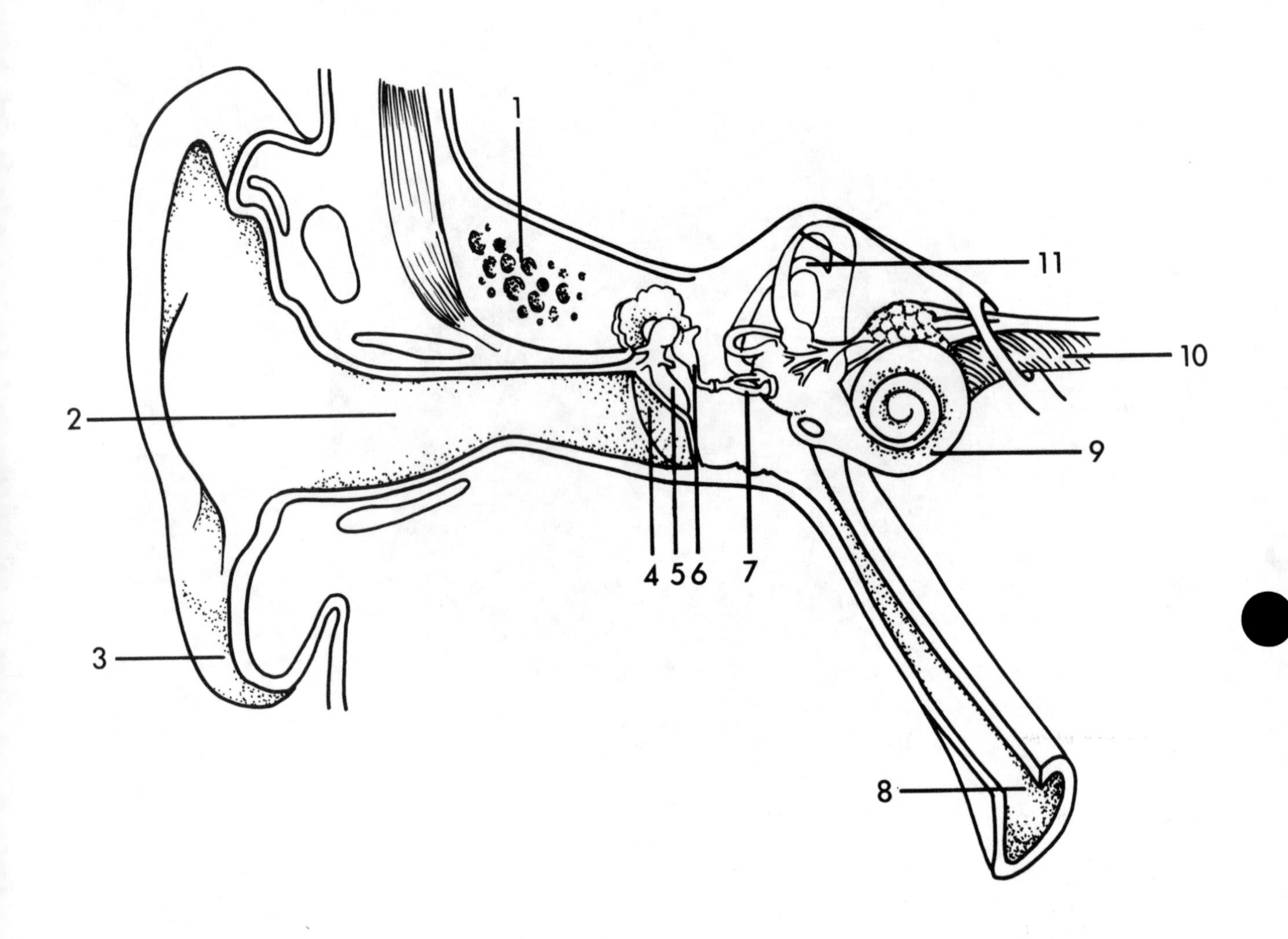

1. temporal bone
2. external auditory canal
3. auricle (pinna)
4. eardrum
5. malleus
6. incus
7. stapes
8. auditory eustachian tube
9. cochlea
10. cochlear nerve
11. semicircular canals

CHAPTER 9

The Endocrine System

The endocrine system has often been compared to a fine concert symphony. When all instruments are playing properly, the sound is melodious. If one instrument plays too loud or too soft, however, it affects the overall quality of the entire performance.

The endocrine system is a ductless system that releases hormones into the bloodstream to help regulate body functions. The pituitary gland may be considered the conductor of the orchestra, as it stimulates many of the endocrine glands to secrete their powerful hormones. All hormones, whether stimulated in this manner or by other control mechanisms, are interdependent. A change in the level of one hormone may affect the level of many other hormones.

In addition to the endocrine glands, prostaglandins, or "tissue hormones," are powerful substances similar to hormones that have been found in a variety of body tissues. These hormones are often produced in a tissue and diffuse only a short distance to act on cells within that area. Prostaglandins influence respiration, blood pressure, gastrointestinal secretions, and the reproductive system and may some day play an important role in the treatment of diseases such as hypertension, asthma, and ulcers.

The endocrine system is a system of communication and control. It differs from the nervous system in that hormones provide a slower, longer lasting effect than do nerve stimuli and responses. Your understanding of the "system of hormones" will alert you to the mechanism of our emotions, response to stress, growth, chemical balances, and many other body functions.

TOPICS FOR REVIEW

Before progressing to Chapter 10, you should be able to identify and locate the primary endocrine glands of the body. Your understanding should include the hormones that are produced by these glands and the method by which these secretions are regulated. Your study will conclude with the pathological conditions that result from the malfunctioning of this system.

MECHANISMS OF HORMONE ACTION
REGULATION OF HORMONE SECRETION
PROSTAGLANDINS

Match the term on the left with the proper selection on the right.

Group A

____1. Pituitary	a. Pelvic cavity
____2. Parathyroids	b. Mediastinum
____3. Adrenals	c. Neck
____4. Ovaries	d. Cranial cavity
____5. Thymus	e. Abdominal cavity

Group B

_____6. Negative feedback
_____7. Tissue hormones
_____8. Second messenger hypothesis
_____9. Exocrine glands
_____10. Target organ cells

a. Explanation for hormone organ recognition
b. Respond to a particular hormone
c. Prostaglandins
d. Discharge secretions into ducts
e. Specialized homeostatic mechanism that regulates release of hormones

Fill in the blanks.

The (11) _______________ _______________ hypothesis is a theory that attempts to explain why hormones cause specific effects in target organs but do not (12) _______________ or act on other organs of the body. Hormones serve as (13) _______________ _______________, providing communication between endocrine glands and (14) _______________ _______________. The second messenger (15) _______________ provides communication within a hormone's (16) _______________ _______________. (17) _______________ _______________ disrupts the normal negative feedback control of hormones throughout the body, and may result in tissue damage, sterility, mental imbalance, and a host of life-threatening metabolic problems.

▸ If you have had difficulty with this section, review pages 184-186. ◂

PITUITARY GLAND
HYPOTHALAMUS

Circle the correct choice.

18. The pituitary gland lies in the _______________ bone.

a. Ethmoid
b. Sphenoid
c. Temporal
d. Frontal
e. Occipital

19. Which one of the following structures would *not* be stimulated by a trophic hormone from the anterior pituitary?

a. Ovaries
b. Testes
c. Thyroid
d. Adrenals
e. Uterus

20. Which one of the following is *not* a function of FSH?

   a. Stimulates the growth of follicles
   b. Stimulates the production of estrogens
   c. Stimulates the growth of seminiferous tubules
   d. Stimulates the interstitial cells of the testes

21. Which one of the following is *not* a function of LH?

   a. Stimulates maturation of a developing follicle
   b. Stimulates the production of estrogens
   c. Stimulates the formation of a corpus luteum
   d. Stimulates sperm cells to mature in the male
   e. Causes ovulation to occur

22. Which one of the following is *not* a function of GH?

   a. Increases glucose catabolism
   b. Increases fat catabolism
   c. Speeds up the movement of amino acids into cells from the bloodstream.
   d. All of the above are functions of GH

23. Which one of the following hormones is *not* released by the anterior pituitary gland?

   a. ACTH
   b. TSH
   c. ADH
   d. FSH
   e. LH

24. Which one of the following is *not* a function of prolactin?

   a. Stimulates breast development during pregnancy
   b. Stimulates milk secretion after delivery
   c. Causes the release of milk from glandular cells of the breast
   d. All of the above are functions of prolactin

25. The anterior pituitary gland:

   a. Secretes eight major hormones
   b. Secretes trophic hormones that stimulate other endocrine glands to grow and secrete
   c. Secretes ADH
   d. Secretes oxytocin

26. TSH acts on the:

   a. Thyroid
   b. Thymus
   c. Pineal
   d. Testes

27. ACTH stimulates the:

   a. Adrenal cortex
   b. Adrenal medulla
   c. Hypothalamus
   d. Ovaries

28. Which hormone is secreted by the posterior pituitary gland?

a. MSH
b. LH
c. GH
d. ADH

29. ADH serves the body by:

a. Initiating labor
b. Accelerating water reabsorption from urine into the blood
c. Stimulating the pineal gland
d. Regulating the calcium/phosphorus levels in the blood

30. The disease caused by hyposecretion of the ADH is:

a. Diabetes insipidus
b. Diabetes mellitus
c. Acromegaly
d. Myxedema

31. The actual production of ADH and oxytocin takes place in which area?

a. Anterior pituitary
b. Posterior pituitary
c. Hypothalamus
d. Pineal

32. Inhibiting hormones are produced by the:

a. Anterior pituitary
b. Posterior pituitary
c. Hypothalamus
d. Pineal

Select the best answer from the choices given and insert the letter in the answer blank.

(a) Anterior pituitary (b) Posterior pituitary (c) Hypothalamus

_____33. Adenohypophysis
_____34. Neurohypophysis
_____35. Induced labor
_____36. Appetite
_____37. Acromegaly
_____38. Body temperature
_____39. Sex hormones
_____40. Trophic hormones
_____41. Gigantism
_____42. Releasing hormones

▸ If you have had difficulty with this section, review pages 190-193. ◂

THYROID GLAND
PARATHYROID GLANDS

Circle the correct answer.

43. The thyroid gland lies (above or below) the larynx.

44. The thyroid gland secretes (calcitonin or glucagon).

45. For thyroxine to be produced in adequate amounts, the diet must contain sufficient (calcium or iodine).

46. Most endocrine glands (do or do not) store their hormones.

47. Colloid is a storage medium for the (thyroid hormone or parathyroid hormone).

48. Calcitonin (increases or decreases) the concentration of calcium in the blood.

49. Simple goiter results from (hyperthyroidism or hypothyroidism).

50. Hyposecretion of thyroid hormones during the formative years leads to (cretinism or myxedema).

51. The parathyroid glands secrete the hormone (PTH or PTA).

52. Parathyroid hormone tends to (increase or decrease) the concentration of calcium in the blood.

▸ If you have had difficulty with this section, review pages 193-195. ◂

ADRENAL GLANDS

Fill in the blanks.

53. The adrenal gland is actually two separate endocrine glands, the ____________ ____________ and the ____________ ____________.

54. Hormones secreted by the adrenal cortex are known as ________________.

55. The outer zone of the adrenal cortex, the zona glomerulosa, secretes ________________.

56. The middle zone, the zona fasciculata, secretes ________________.

57. The innermost zone, the zona reticularis, secretes ________________.

58. Glucocorticoids act in several ways to increase ________________.

59. Glucocorticoids also play an essential part in maintaining ____________ ____________.

60. The adrenal medulla secretes the hormones ________________ and ________________.

61. The adrenal medulla may help the body resist ________________.

62. The term ____________ ____________ ____________ is often used to describe how the body mobilizes a number of different defense mechanisms when threatened by harmful stimuli.

Select the best response from the choices given and insert the letter in the answer blank.

(a) Adrenal cortex (b) Adrenal medulla

____63. Addison's disease
____64. Anti-immunity
____65. Adrenaline

____66. Cushing's syndrome
____67. Fight or flight syndrome
____68. Aldosterone
____69. Androgens

▸ If you have had difficulty with this section, review pages 196-200. ◂

PANCREATIC ISLETS
SEX GLANDS
THYMUS
PLACENTA
PINEAL GLAND

Circle the term that does not belong.

70. Alpha cells	Glucagon	Beta cells	Glycogenolysis
71. Insulin	Glucagon	Beta cells	Diabetes mellitus
72. Estrogens	Progesterone	Corpus luteum	Thymosin
73. Chorion	Interstitial cells	Testosterone	Semen
74. Immune system	Mediastinum	Aldosterone	Thymosin
75. Pregnancy	ACTH	Estrogen	Chorion
76. Melatonin	Menstruation	"Third eye"	Semen

▸ If you have had difficulty with this section, review pages 201-203. ◂

APPLYING WHAT YOU KNOW

77. Mrs. Fortner made a routine visit to her physician last week. When the laboratory results came back, the report indicated a high level of chorionic gonadotropin in her urine. What did this mean to Mrs. Fortner?

78. Mrs. Calhoun noticed that her daughter was beginning to take on the secondary sex characteristics of a male. The pediatrician diagnosed the condition as a tumor of an endocrine gland. Where specifically was the tumor located?

79. WORD FIND

Can you find the terms from the chapter listed below in the box of letters? Words may be spelled top to bottom, bottom to top, right to left, left to right, or diagonally.

```
J C W W H Y P E R C A L C E M I A L E D
T P Y E S K A A R T E S Y E Z S G Z Z E
B O U D D O M L E T X D J N I Y D B P O
L F Y X D E S W T G W I R O F A G D M S
N H S S D Q P F I U P A G M U V E Q G Z
O Q Z E Q N U A O I M B B R T J F M W M
P V X U V F O D G Z A E K O L O R Y G V
R Y G R Z O H G X T O T G H V K E D T E
M J I O I X D J A A N E D E F D N J Q H
N V J Q V M H Y Y C O S C Q I N D Z P Y
C R E T I N I S M Q U L P U N L O I B P
E X O C R I N E Z Z M L R K N R C Y S O
V D A R I B T J V P C E G L C U R F I G
S E H R I T Q O K J S F G W Q K I T M L
R C O H J F R Q S I M F D I X P N L E Y
X R O M T L L P S U E B X D E M E T O C
M F S D I O C I T R O C N Z P O Q K J E
M S T R E S S S D I O R E T S M T A M M
U P P U N O I T A Z I N I E T U L H V I
U P R O S T A G L A N D I N S C E A C A
```

LIST OF WORDS

CORTICOIDS	CRETINISM	DIABETES
DIURESIS	ENDOCRINE	EXOCRINE
GOITER	HORMONE	HYPERCALCEMIA
HYPOGLYCEMIA	LUTEINIZATION	STEROIDS
MYXEDEMA	PROSTAGLANDINS	STRESS
GLUCAGON		

DID YOU KNOW?

The total daily output of the pituitary gland is less than 1/1,000,000 of a gram.

THE ENDOCRINE SYSTEM

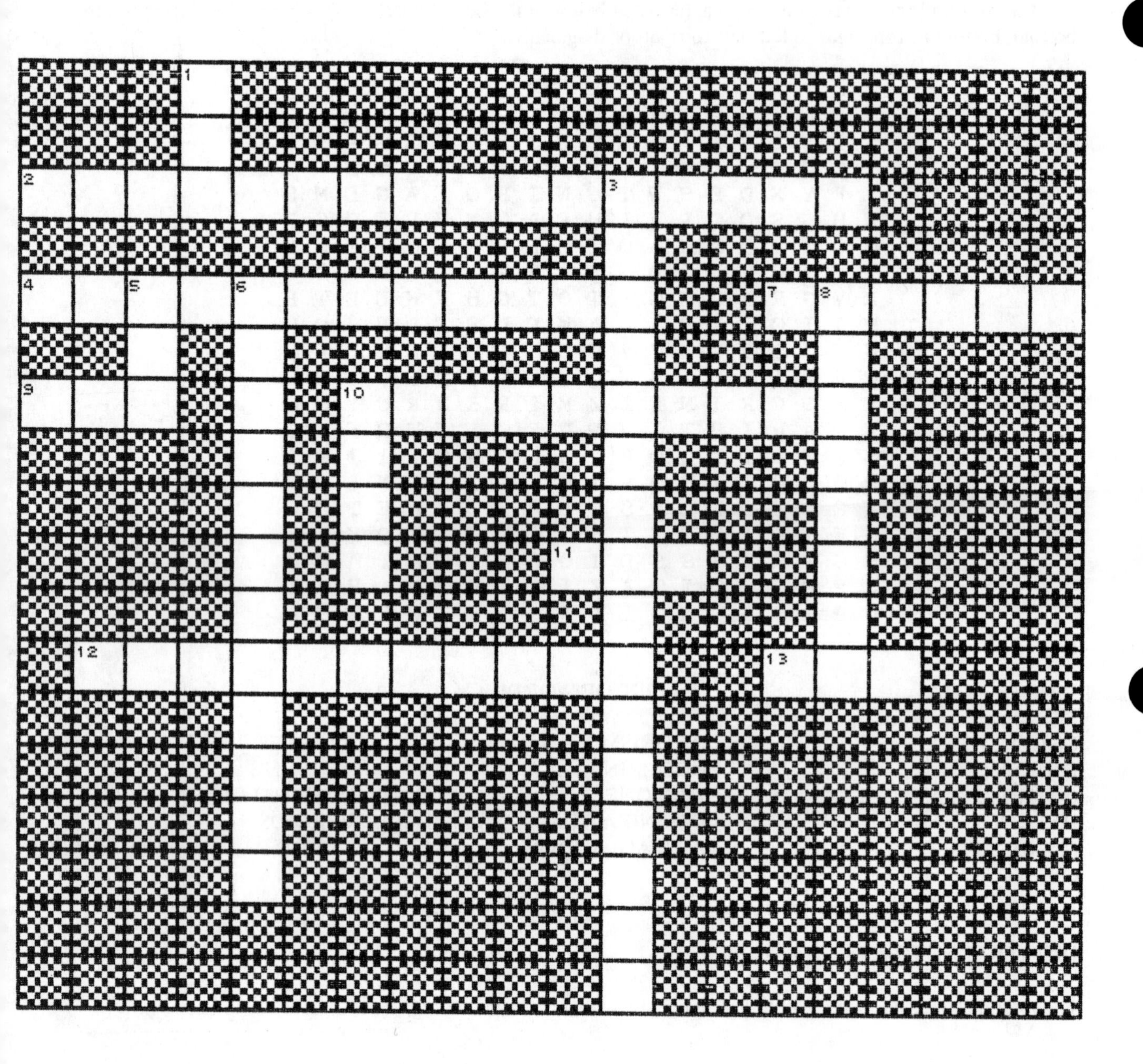

ACROSS

2. Hypersecretion of glucocorticoids
4. Hypersecretion of insulin
7. Hyposecretion of thyroid
9. Antagonist to diuresis
10. Hypersecretion of growth hormone
11. Estrogens
12. Adrenal medulla
13. Secreted by cells in the walls of the heart's atria

DOWN

1. Melanin
3. Hyposecretion of Islands of Langerhans (2 words)
5. Increases calcium concentration
6. Converts amino acids to glucose
8. Labor
10. Adrenal cortex

ENDOCRINE GLANDS

1. ____________________
2. ____________________
3. ____________________
4. ____________________
5. ____________________
6. ____________________
7. ____________________
8. ____________________
9. ____________________

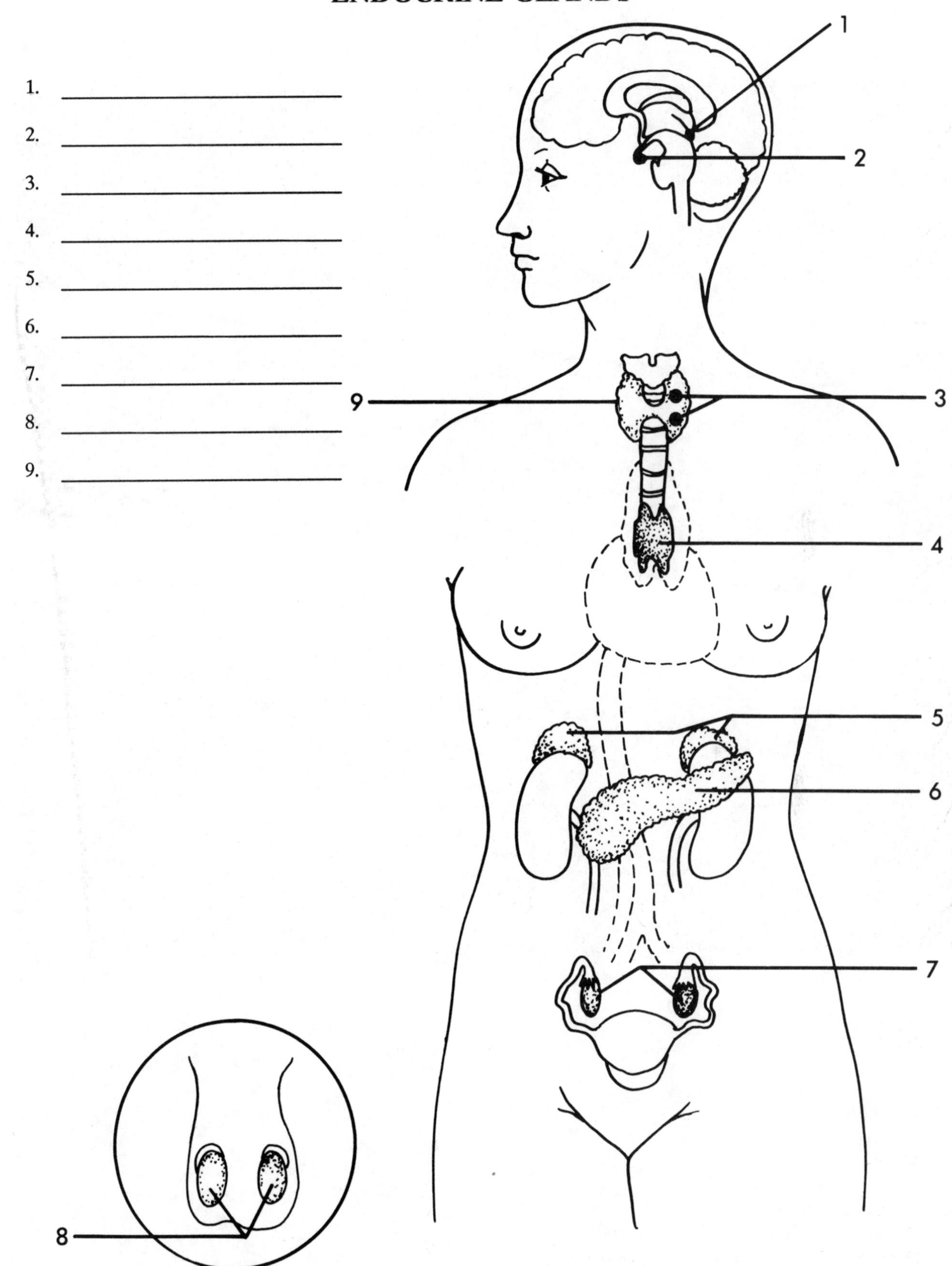

CHAPTER 10

Blood

Blood, the river of life, is the body's primary means of transportation. Although it is the respiratory system that provides oxygen for the body, the digestive system that provides nutrients, and the urinary system that eliminates wastes, none of these functions could be provided for the individual cells without the blood. In less than one minute, a drop of blood will complete a trip through the entire body, distributing nutrients and collecting the wastes of metabolism.

Blood is divided into plasma, which is the liquid portion of blood, and the formed elements, which are the blood cells. There are three types of blood cells: red blood cells, white blood cells, and platelets. Together these cells and plasma provide a means of transportation that delivers the body's daily necessities.

Although the red blood cells in all of us are of a similar shape, we have different blood types. Blood types are identified by the presence of certain antigens in the red blood cells. Every person's blood belongs to one of four main blood groups: Type A, B, AB, or O. Any one of the four groups or "types" may or may not have the Rh factor present in the red blood cells. If an individual has a specific antigen called the Rh factor present in his or her blood, the blood is Rh positive. If this factor is missing, the blood is Rh negative. Approximately, 85% of the population have the Rh factor (Rh positive) while 15% do not have the Rh factor (Rh negative).

Your understanding of this chapter will be necessary to prepare a proper foundation for the circulatory system.

TOPICS FOR REVIEW

Before progressing to Chapter 11, you should have an understanding of the structure and function of blood plasma and cells. Your review should also include a knowledge of blood types and Rh factors.

BLOOD COMPOSITION

Circle the best answer.

1. Another name for white blood cells is:

   a. Erythrocytes
   b. Leukocytes
   c. Thrombocytes
   d. Platelets

2. Another name for platelets is:

   a. Neutrophils
   b. Eosinophils
   c. Thrombocytes
   d. Erythrocytes

3. Pernicious anemia is caused by:

   a. A lack of vitamin B_{12}
   b. Hemorrhage
   c. Radiation
   d. Bleeding ulcers

4. The laboratory test called hematocrit tells the physician:

   a. The volume of white cells in a blood sample
   b. The volume of red cells in a blood sample
   c. The volume of platelets in a blood sample
   d. The volume of plasma in a blood sample

5. An example of a nongranular leukocyte is a/an:

   a. Platelet
   b. Erythrocyte
   c. Eosinophil
   d. Monocyte

6. An abnormally high white blood cell count is known as:

   a. Leukemia
   b. Leukopenia
   c. Leukocytosis
   d. Anemia

7. A critical component of hemoglobin is:

   a. Potassium
   b. Calcium
   c. Vitamin K
   d. Iron

8. Sickle cell anemia is caused by:

   a. The production of an abnormal type of hemoglobin
   b. The production of excessive neutrophils
   c. The production of excessive platelets
   d. The production of abnormal leukocytes

9. The practice of using blood transfusions to increase oxygen delivery to muscles during athletic events is called:

   a. Blood antigen
   b. Blood doping
   c. Blood agglutination
   d. Blood proofing

10. The term used to describe the condition of a circulating blood clot is:

    a. Thrombosis
    b. Embolism
    c. Hemoglobin
    d. Platelet

11. Which one of the following types of cells is *not* a granular leukocyte?

    a. Neutrophil
    b. Monocyte
    c. Basophil
    d. Eosinophil

12. If a blood cell has no nucleus and is shaped like a biconcave disc, then the cell most likely is a/an:

    a. Platelet
    b. Lymphocyte
    c. Basophil
    d. Eosinophil
    e. Red blood cell

13. Red bone marrow forms all kinds of blood cells except:

a. Platelets
(b.) Lymphocytes
c. Red blood cells
d. Neutrophils

14. Myeloid tissue is found in all but which one of the following locations?

a. Sternum
b. Ribs
(c.) Wrist bones
d. Hip bones
e. Cranial bones

15. Lymphatic tissue is found in all but which of the following locations?

a. Lymph nodes
b. Thymus
c. Spleen
(d.) All of the above contain lymphatic tissue

16. The "buffy coat" layer in a hematocrit tube contains:

a. Red blood cells
b. Plasma
c. Platelets
d. White blood cells
(e.) Two of the above are correct

17. The hematocrit value for red blood cells is ____________%.

a. 75
b. 60
c. 50
(d.) 45
e. 35

18. An unusually low white blood cell count would be termed:

a. Leukemia
(b.) Leukopenia
c. Leukocytosis
d. Anemia
e. None of the above is correct

19. Most of the oxygen transported in the blood is carried by:

a. Platelets
b. Plasma
c. White blood cells
(d.) Red blood cells
e. None of the above is correct

20. The most numerous of the phagocytes are the ____________________.

a. Lymphocytes
(b.) Neutrophils
c. Basophils
d. Eosinophils
e. Monocytes

21. Which one of the following types of cells is *not* phagocytic?

a. Neutrophils
b. Eosinophils
(c.) Lymphocytes
d. Monocytes
e. All of the above are phagocytic cells

22. Which of the following cell types functions in the immune process?

a. Neutrophils
b. Lymphocytes
c. Monocytes
d. Basophils
e. Reticuloendothelial cells

23. The organ that manufactures prothrombin is the:

a. Liver
b. Pancreas
c. Thymus
d. Kidney
e. Spleen

24. Which one of the following vitamins acts to accelerate blood clotting?

a. A
b. B
c. C
d. D
e. K

25. Which one of the following substances is *not* a part of the plasma?

a. Hormones
b. Salts
c. Nutrients
d. Wastes
e. All of the above are part of the plasma

26. The normal volume of blood in an adult is about:

a. 2-3 pints
b. 2-3 quarts
c. 2-3 gallons
d. 4-6 liters

27. Blood is normally:

a. Very acidic
b. Slightly acidic
c. Neutral
d. Slightly alkaline

▸ If you have had difficulty with this section, review pages 210-218. ◂

BLOOD TYPES
RH FACTOR

Fill in the blank areas.

28.

Blood Type	Antigen Present in RBC	Antibody Present in Plasma
A	A	Anti-B
B	B	Anti-A
AB	A+B	None
O	None	Anti-A+B

Fill in the blanks

29. An antigen is a substance that can stimulate the blood to make antibodies.
30. An antibody is a substance made by the body in response to stimulation by an antigen.
31. Many antibodies react with their antigens to clump or ________________ them.
32. If a baby is born to an Rh-negative mother and Rh-positive father, it may develop the disease erythroblastosis fetalis.
33. The term "Rh" is used because the antigen was first discovered in the blood of a rhesus monkey.

► If you have had difficulty with this section, review pages 217-218. ◄

APPLYING WHAT YOU KNOW

34. Mrs. Payne's blood type is O positive. Her husband's type is O negative. Her newborn baby's blood type is O negative. Is there any need for concern with this combination?
35. After Mrs. Freund's baby was born, the doctor applied a gauze dressing for a short time on the umbilical cord. He also gave the baby a dose of vitamin K. Why did the doctor perform these two procedures?

36. **WORD FIND**

Can you find the terms from this chapter listed below in the box of letters? Words may be spelled top to bottom, bottom to top, right to left, left to right, or diagonally.

```
I T C S U B M O R H T E L Z F Q I H T Q
X W R Y B H E Y P H A G O C Y T E S L N
G Q E K A E M X M L T R P I N J X V E H
L V C Q S M B R N M T H R O M B I N U V
E W I L O O O U A Q K O H T S U M I K S
U O P N P G L Q P Q R Y X E S O H S E Z
K M I M H L U C A O M X T F N B W R M V
O G E D I O S O I Y P Y A O R G S J I Z
C S N N L B J I D Q C C C X Z J I E A H
Y U T M P I L V S O T Y F J W E J Q U S
T S T R O N O D R O T N A E O A E B C W
E E X P B Z F H R E Y N T V W E U S M I
S H N V K G T D H H T A F G Q M B I A T
S R I M J Y L T T I X N A M S A L P Y T
F R R T R W O C G I G T I E P Y T A B S
U Q B E I P L E V S N I B C L Y U E E R
V T I B S I N R X W O B X Y C K E R R C
E Z F E N A I M E N A O J S E K U X H S
N I R A P E H O X Z I D P M Z M F D Z Q
H E M A T O C R I T R Y Z I P N C L G W
```

LIST OF WORDS

PLASMA
ANEMIA
AIDS
MONOCYTE
HEPARIN
FACTOR
ANTIGEN
DONOR

ERYTHROCYTES
HEMOGLOBIN
LEUKEMIA
THROMBUS
EMBOLUS
TYPE
THROMBIN
RECIPIENT

LEUKOCYTES
HEMATOCRIT
PHAGOCYTES
FIBRIN
SERUM
ANTIBODY
BASOPHIL
RHESUS

DID YOU KNOW?

Every pound of excess fat contains some 200 miles of additional capillaries to push blood through.

BLOOD

LEUKOCYTOSIS
FIBRIN
ANTIGENS
THROMBOCYTES
PHAGOCYTOSIS
ERYTHROCYTE
PLASMA
THROMBUS

UNIVERSAL DONOR
EOSINOPHILS
ANEMIA
EMBOLISM
HEMOGLOBIN
HEPARIN

ACROSS

1. Abnormally high WBC count
4. Final stage of clotting process
5. Substances that stimulate the body to make antibodies
7. Platelets
8. To engulf and digest microbes
11. RBC
12. Liquid portion of blood
13. Stationary blood clot

DOWN

2. Type O (two words)
3. Type of leukocyte
5. Inability of the blood to carry sufficient O2
6. Circulating blood clot
9. O2 carrying mechanism of blood
10. Prevents clotting of blood

HUMAN BLOOD CELLS

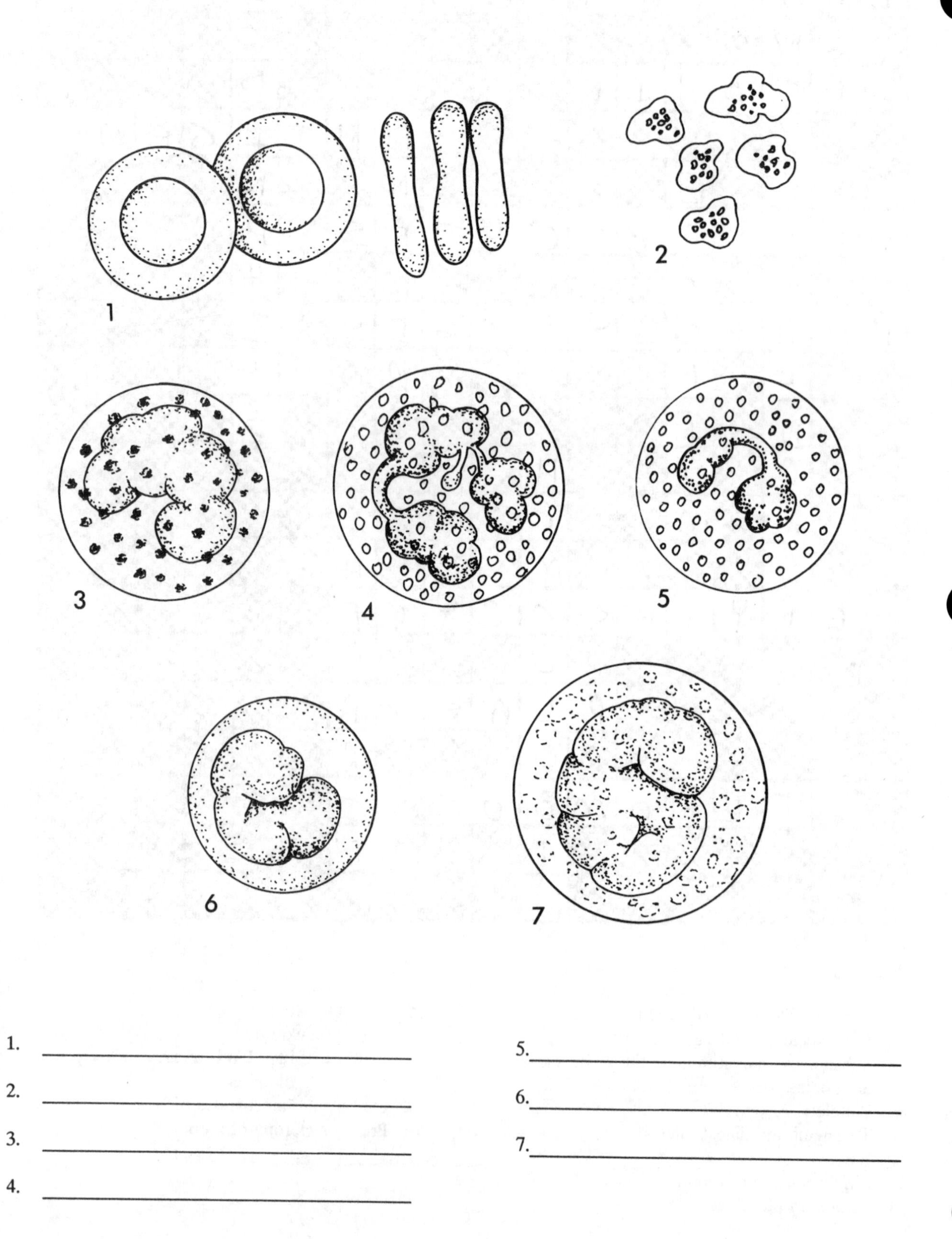

1. ______________________________
2. ______________________________
3. ______________________________
4. ______________________________
5. ______________________________
6. ______________________________
7. ______________________________

BLOOD TYPING

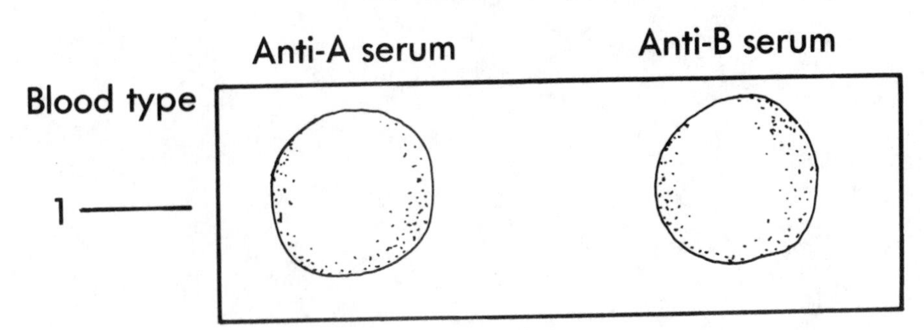

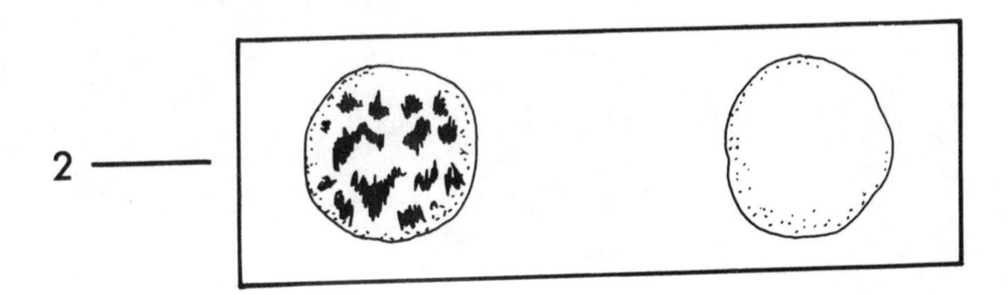

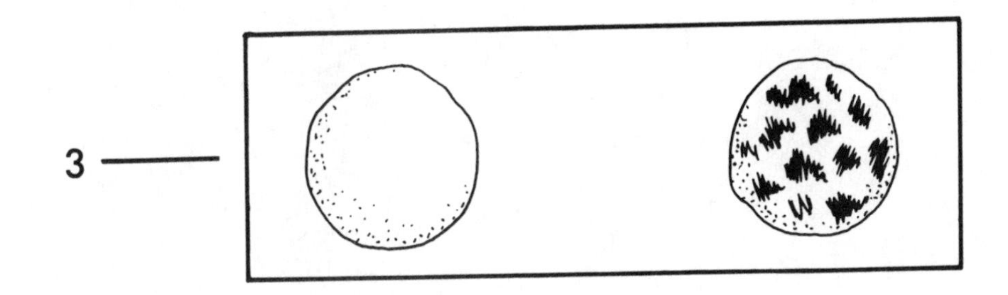

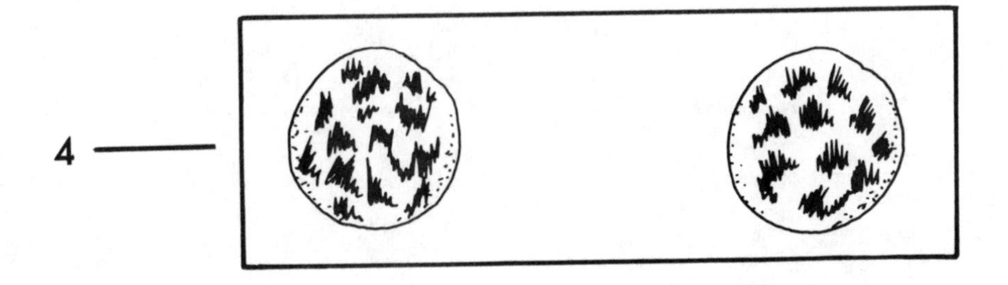

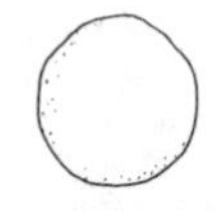

Normal blood

Agglutinated blood

CHAPTER 11

Circulatory System

The heart is actually two pumps, one to move blood to the lungs, the other to push it out into the body. These two functions seem rather elementary by comparison to the complex and numerous functions performed by most of the other body organs, and yet, if this pump stops, within a few short minutes all life ceases.

The heart is divided into two upper compartments called atria or receiving chambers and two lower compartments or discharging chambers called ventricles. By age 45, approximately 300,000 tons of blood will have passed through these chambers to be circulated to the blood vessels. These vessels, called arteries, veins, and capillaries serve different functions. Arteries carry blood from the heart, veins carry blood to the heart, and capillaries are exchange vessels or connecting links between the arteries and veins. This closed system of circulation provides distribution of blood to the whole body (systemic circulation) and to specific regions, such as pulmonary circulation or hepatic portal circulation.

Blood pressure is the force of blood in the vessels. This force is highest in arteries and lowest in veins. Normal blood pressure varies among individuals and depends on the volume of blood in the arteries. The larger the volume of blood in the arteries, the more pressure is exerted on the walls of the arteries, and the higher the arterial pressure. Conversely, the less blood in the arteries, the lower the blood pressure.

A functional cardiovascular system is vital for survival because without circulation, tissues would lack a supply of oxygen and nutrients. Waste products would begin to accumulate and could become toxic. Your review of this system will provide you with an understanding of the complex transportation mechanism of the body necessary for survival.

TOPICS FOR REVIEW

Before progressing to Chapter 12 you should have an understanding of the structure and function of the heart and blood vessels. Your review should include a study of systemic, pulmonary, hepatic portal, and fetal circulations, and should conclude with a thorough understanding of blood pressure and pulse.

HEART

Fill in the blanks.

1. Rhythmic compression of the heart combined with effective artificial respiration is known as CPR.
2. The septum divides the heart into right and left sides between the atria.
3. The atria are the two upper chambers of the heart.
4. The ventricles are the two lower chambers of the heart.
5. The cardiac muscle tissue is referred to as the myocardium
6. Inflammation of the heart lining is Endocarditis.

7. The two AV valves are bicuspid and tricuspid.

8. Pulmonary Circulation involves movement of blood from the right ventricle to the lungs.

9. An occlusion of a coronary artery is known as Pulmonary Circulation

10. Myocardial infarction occurs when heart muscle cells are deprived of oxygen and become damaged or die.

11. The pacemaker of the heart is the Sinoatrial node.

12. A normal ECG tracing has three characteristic waves. They are __________, __________, and __________ waves.

13. Repolarization begins just before the relaxation phase of cardiac muscle activity noted on an ECG.

Select the best answer.

a. Pericardium
b. Severe chest pain
c. Thrombosis
d. Pulmonary
e. Heart block
f. Ventricles
g. Systemic
h. Coronary arteries
i. Systole
j. Depolarization
k. Atria
l. Apex
m. Epicardium

a 14. Covering of heart
K 15. Receiving chambers
g 16. Circulation from left ventricle throughout body
c 17. Blood clot
d 18. Semilunar valve
f 19. Discharging chambers
h 20. Supplies oxygen to heart muscle
b 21. Angina pectoris
e 22. Slow heart rate caused by blocked impulses
l 23. Contraction of the heart
j 24. Electrical activity associated with ECG
i 25. Blunt-pointed lower edge of heart
m 26. Visceral pericardium

▸ If you have had difficulty with this section, review pages 224-233. ◂

BLOOD VESSELS
CIRCULATION

Match the term on the left with the proper selection on the right.

d 27. Arteries
b 28. Veins
c 29. Capillaries
g 30. Tunica externa
a 31. Precapillary sphincters
e 32. Superior vena cava
f 33. Aorta

a. Smooth muscle cells that guard entrance to capillaries
b. Carry blood to the heart
c. Carry blood into venules
d. Carry blood away from the heart
e. Largest vein
f. Largest artery
g. Outermost layer of arteries and veins

Circle the best answer.

34. The aorta carries blood out of the:

a. Right atrium
b. Left atrium
c. Right ventricle
d. Left ventricle
e. None of the above is correct

35. The superior vena cava returns blood to the:

a. Left atrium
b. Left ventricle
c. Right atrium
d. Right ventricle
e. None of the above is correct

36. Which one of the following vessels has its wall made up entirely of endothelial cells?

a. Vein
b. Capillary
c. Artery
d. Venule
e. Arteriole

37. The ______________________ is made up of smooth muscle.

a. Tunica media
b. Tunica adventitia
c. Tunica intima
d. Endothelium
e. Myocardium

38. The ______________ function as exchange vessels.

a. Venules
b. Capillaries
c. Arteries
d. Arterioles
e. Veins

39. Blood returns from the lungs during pulmonary circulation via the:

a. Pulmonary artery
b. Pulmonary veins
c. Aorta
d. Inferior vena cava

40. The hepatic portal circulation serves the body by:

a. Removing excess glucose and storing it in the liver as glycogen
b. Detoxifying blood
c. Removing various poisonous substances present in blood
d. All of the above

41. The structure used to bypass the liver in fetal circulation is the:

a. Foramen ovale
b. Ductus venosus
c. Ductus arteriosus
d. Umbilical vein

42. The foramen ovale serves the fetal circulation by:

a. Connecting the aorta and the pulmonary artery
b. Shunting blood from the right atrium directly into the left atrium
c. Bypassing the liver
d. Bypassing the lungs

43. The structure used to connect the aorta and pulmonary artery in fetal circulation is the:

a. Ductus arteriosus
b. Ductus venosus
c. Aorta
d. Foramen ovale

44. Which of the following is *not* an artery?

a. Femoral
b. Popliteal
c. Coronary
d. Inferior vena cava

45. Which of the following has valves to assist the blood flow:

a. Veins
b. Arteries
c. Capillaries
d. Arterioles

▸ If you have had difficulty with this section, review pages 233-243. ◂

BLOOD PRESSURE
PULSE

Mark T if the answer is true. If the answer is false, circle the wrong word(s) and correct the statement by inserting the proper word(s) in the answer blank.

__________46. Blood pressure is highest in the veins and lowest in the arteries.
__________47. The difference between two blood pressures is referred to as blood pressure deficit.
__________48. If the blood pressure in the arteries were to decrease so that it became equal to the average pressure in the arterioles, circulation would increase.
__________49. A stroke is often the result of low blood pressure.
__________50. Massive hemorrhage increases blood pressure.
__________51. Blood pressure is the volume of blood in the vessels.
__________52. Both the strength and the rate of heartbeat affect cardiac output and blood pressure.
__________53. The diameter of the arterioles helps to determine how much blood drains out of arteries into arterioles.

_________ 54. A stronger heartbeat tends to decrease blood pressure and a weaker heartbeat tends to increase it.

_________ 55. The systolic pressure is the pressure while the ventricles relax.

_________ 56. The diastolic pressure is the pressure while the ventricles contract.

_________ 57. The pulse is a vein expanding and then recoiling.

_________ 58. The radial artery is located at the wrist.

_________ 59. The common carotid artery is located in the neck along the front edge of the sternocleidomastoid muscle.

_________ 60. The artery located at the bend of the elbow and used for locating the pulse is the dorsalis pedis.

▸ If you have had difficulty with this section, review pages 243-246. ◂

Unscramble the words.

61. STMESYCI

62. NULVEE

63. RYTREA

64. USLEP

Take the circled letters, unscramble them, and fill in the statement.

How Noah survived the flood.

65.

APPLYING WHAT YOU KNOW

66. Else was experiencing angina pectoris. Her doctor suggested a surgical procedure that would require the removal of a vein from another region of her body. They would then use the vein to bypass a partial blockage in her coronary arteries. What is this procedure called?

67. Mrs. Frank has heart block. Her electrical impulses are being blocked from reaching the ventricles. An electrical device that causes ventricular contractions at a rate necessary to maintain circulation is being considered as possible treatment for her condition. What is this device?

68. Mrs. Calhoun was diagnosed with an acute case of endocarditis. What is the real danger of this diagnosis?

DID YOU KNOW?

Your heart pumps more than 5 quarts of blood every minute or 2,000 gallons a day.

CIRCULATORY SYSTEM

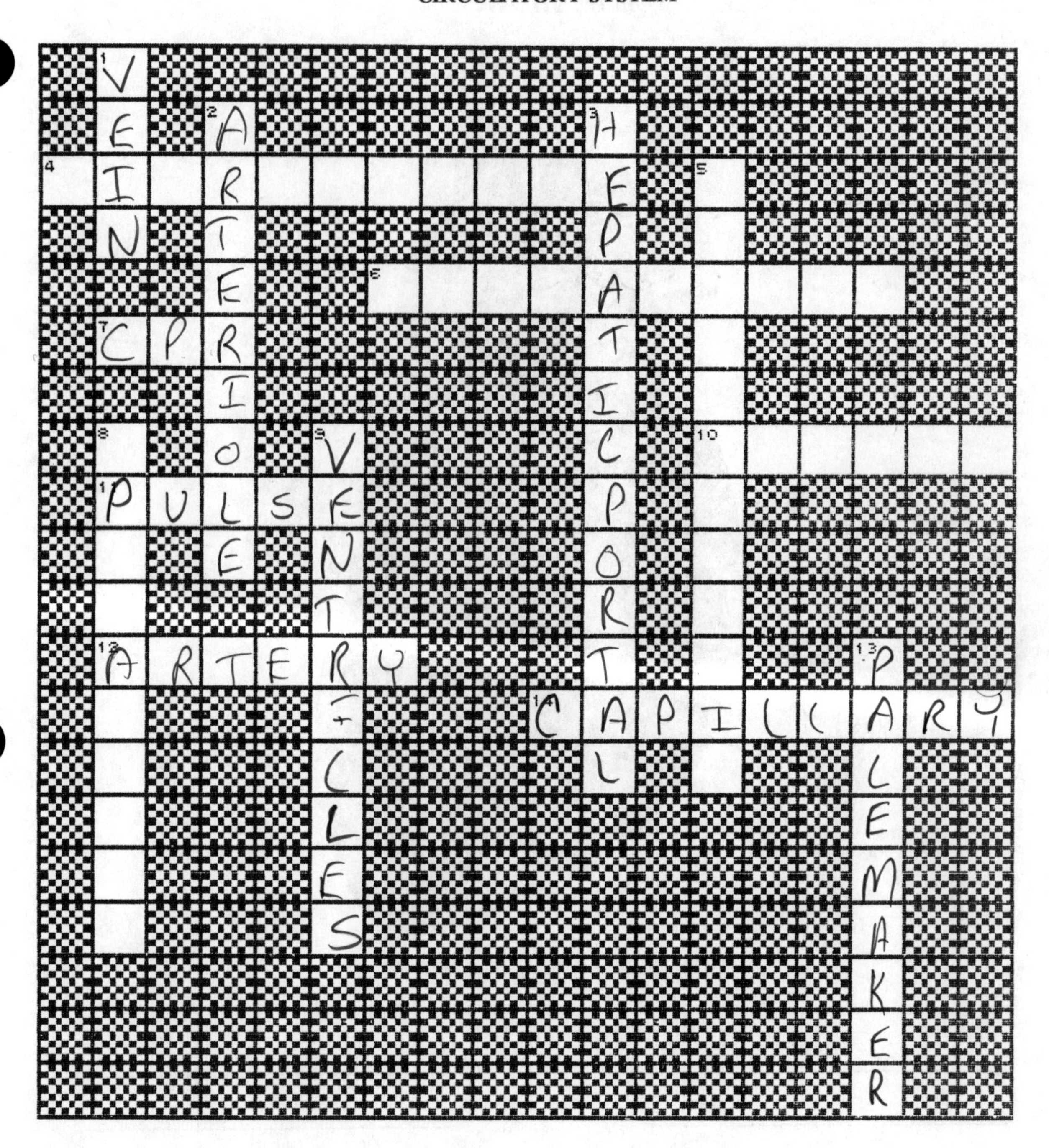

ACROSS

4. Bicuspid valve (2 words)
6. Muscular layer of heart
7. Cardiopulmonary resuscitation (abbrev.)
10. Upper chamber of heart
11. Heart rate
12. Carries blood away from heart
14. Carries blood from arterioles into venules

DOWN

1. Carries blood to the heart
2. Tiny artery
3. Unique blood circulation through the liver (2 words)
5. Inflammation of the lining of the heart
8. Inner layer of pericardium
9. Lower chambers of the heart
13. SA node

THE HEART

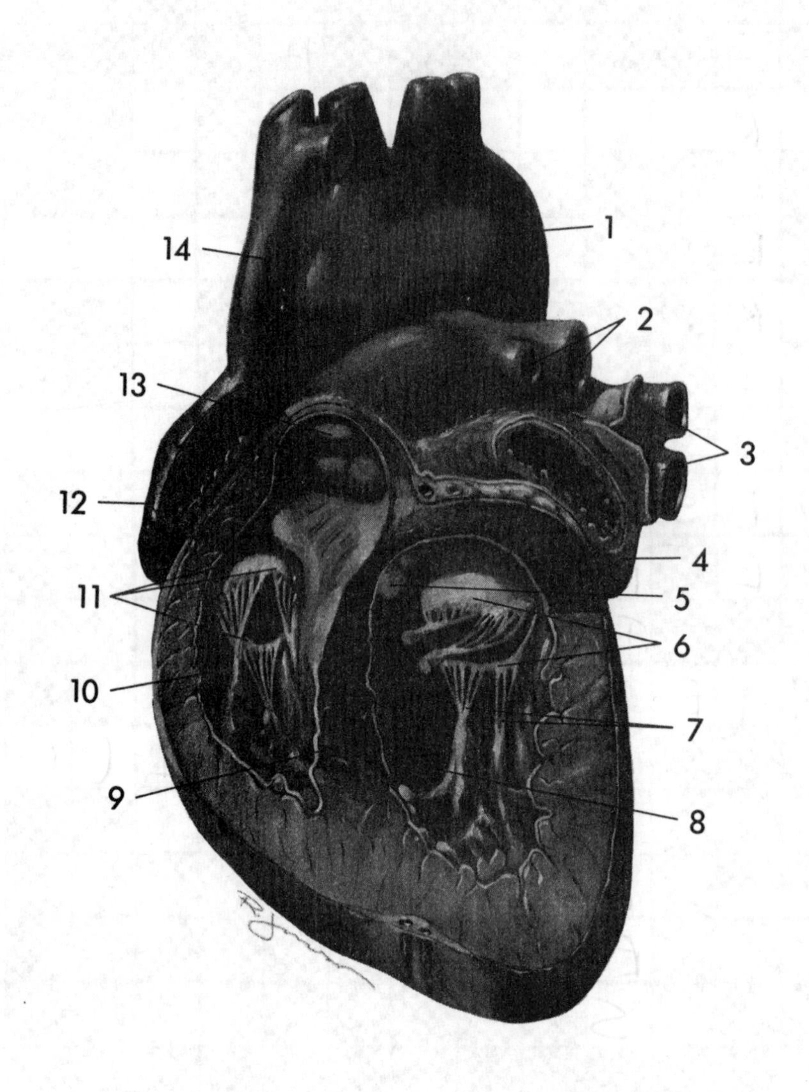

1. ______________________

2. ______________________

3. ______________________

4. ______________________

5. ______________________

6. ______________________

7. ______________________

8. ______________________

9. ______________________

10. ______________________

11. ______________________

12. ______________________

13. ______________________

14. ______________________

CONDUCTION SYSTEM OF THE HEART

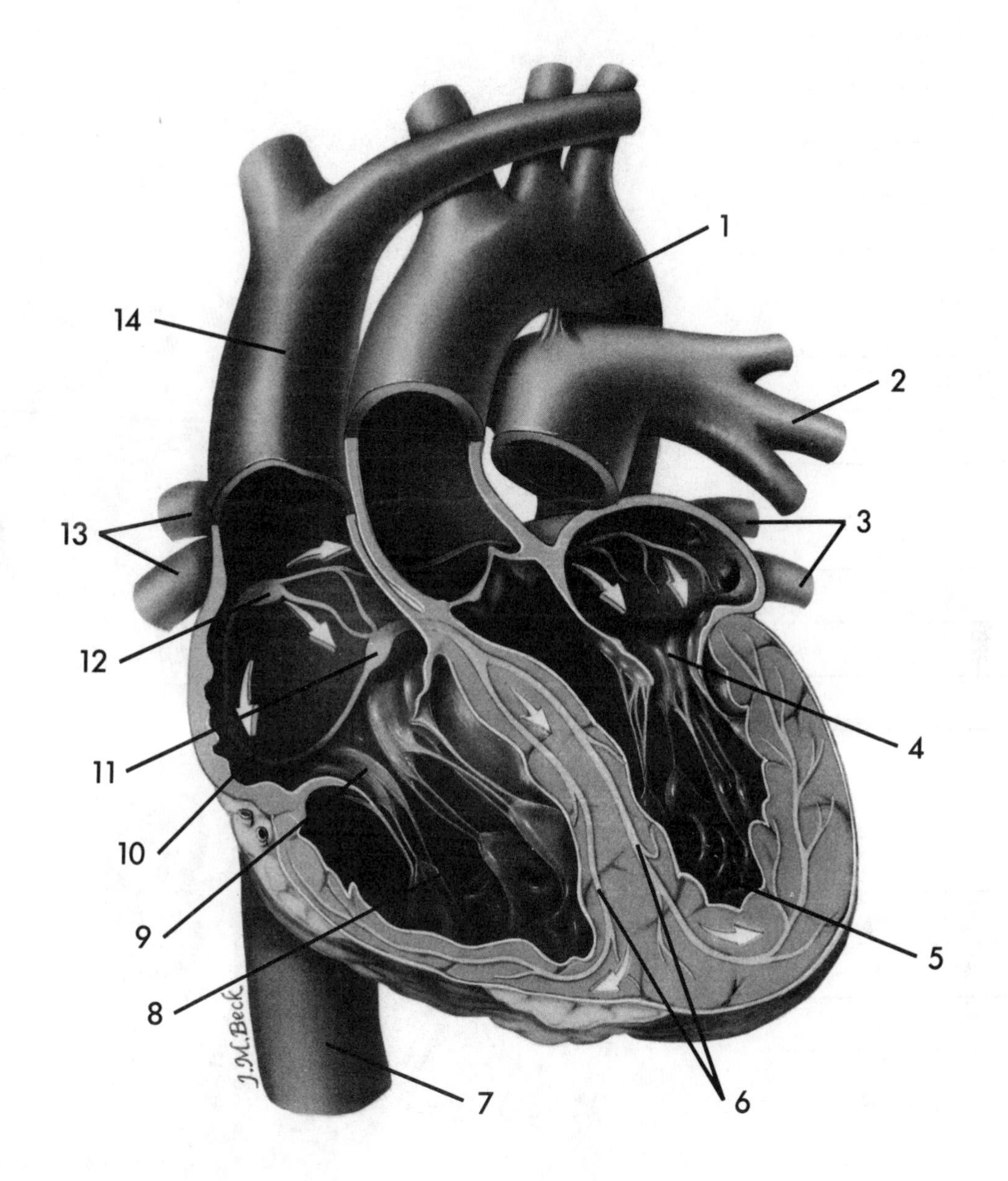

1. ______________________

2. ______________________

3. ______________________

4. ______________________

5. ______________________

6. ______________________

7. ______________________

8. ______________________

9. ______________________

10. ______________________

11. ______________________

12. ______________________

13. ______________________

14. ______________________

FETAL CIRCULATION

1. ______________________
2. ______________________
3. ______________________
4. ______________________
5. ______________________
6. ______________________
7. ______________________
8. ______________________
9. ______________________
10. ______________________
11. ______________________
12. ______________________
13. ______________________
14. ______________________
15. ______________________
16. ______________________
17. ______________________
18. ______________________
19. ______________________
20. ______________________
21. ______________________
22. ______________________
23. ______________________
24. ______________________
25. ______________________
26. ______________________
27. ______________________
28. ______________________

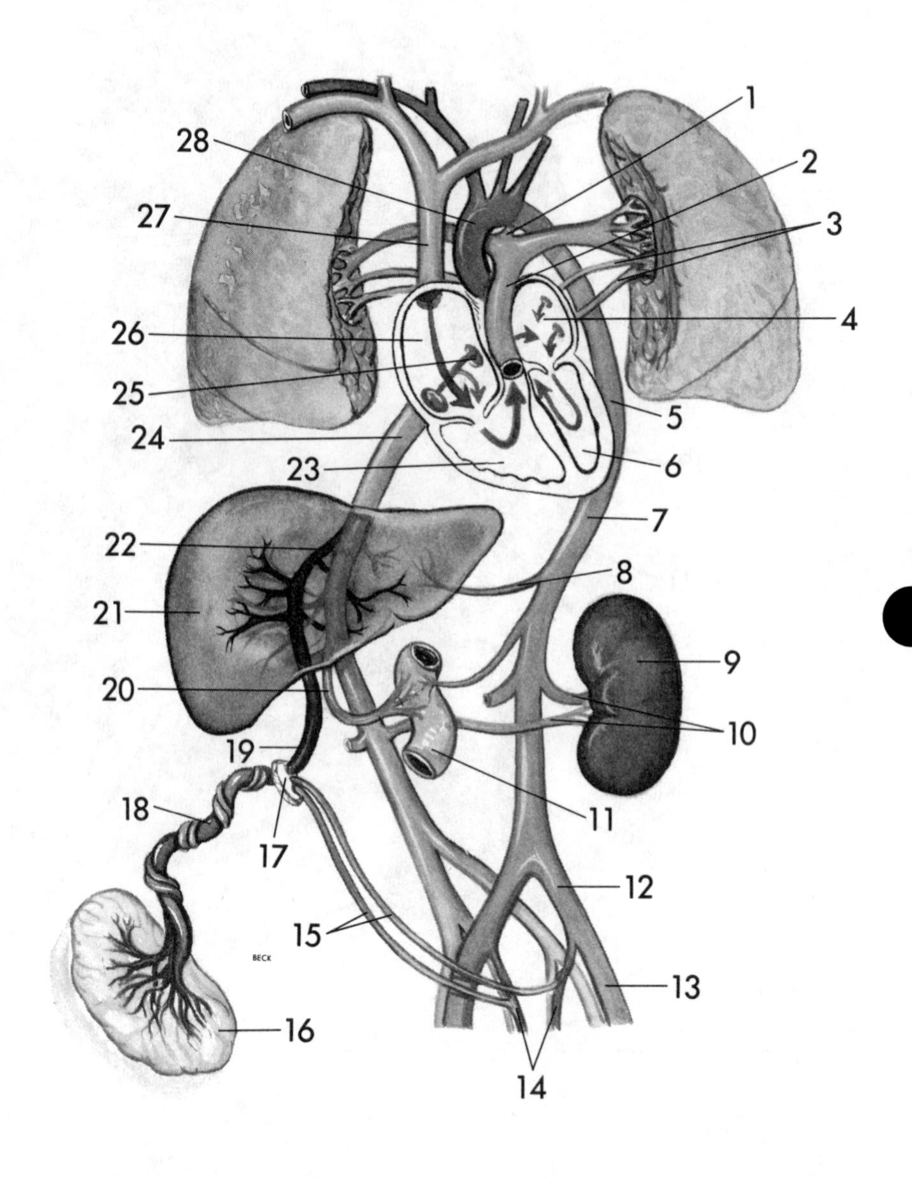

HEPATIC PORTAL CIRCULATION

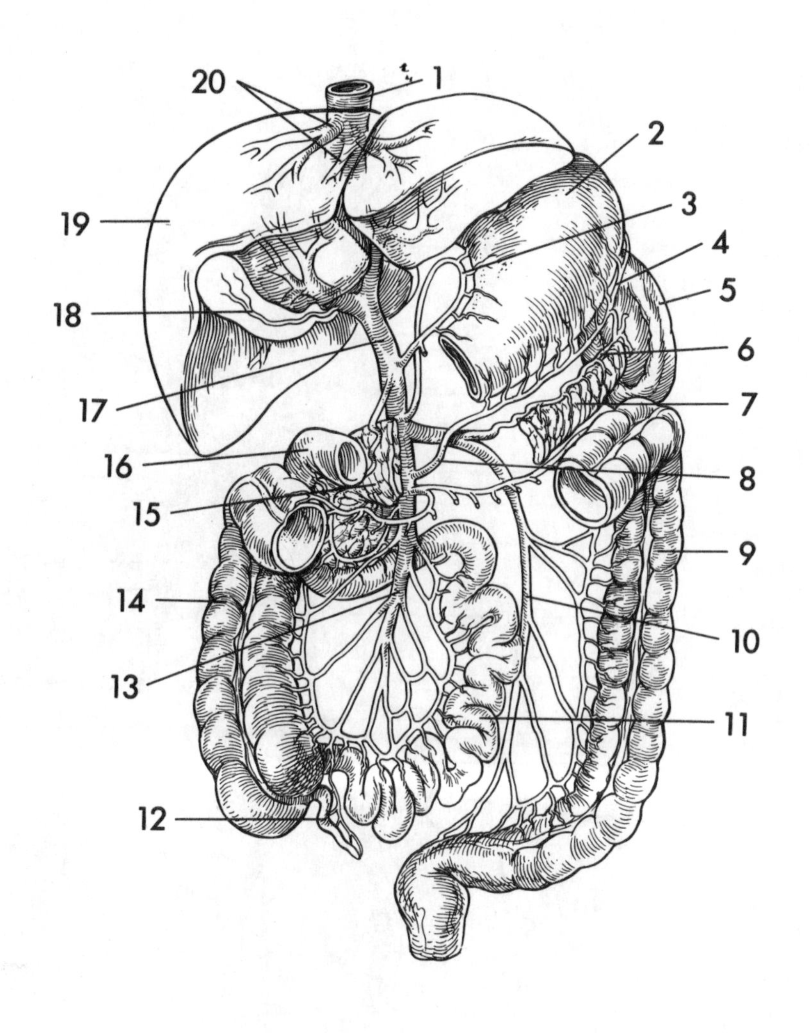

1. ______________________ 11. ______________________

2. ______________________ 12. ______________________

3. ______________________ 13. ______________________

4. ______________________ 14. ______________________

5. ______________________ 15. ______________________

6. ______________________ 16. ______________________

7. ______________________ 17. ______________________

8. ______________________ 18. ______________________

9. ______________________ 19. ______________________

10. ______________________ 20. ______________________

PRINCIPAL ARTERIES OF THE BODY

1. ______________
2. ______________
3. ______________
4. ______________
5. ______________
6. ______________
7. ______________
8. ______________
9. ______________
10. ______________
11. ______________
12. ______________
13. ______________
14. ______________
15. ______________
16. ______________
17. ______________
18. ______________
19. ______________
20. ______________
21. ______________
22. ______________
23. ______________
24. ______________
25. ______________
26. ______________
27. ______________
28. ______________
29. ______________
30. ______________

PRINCIPAL VEINS OF THE BODY

1. ______________
2. ______________
3. ______________
4. ______________
5. ______________
6. ______________
7. ______________
8. ______________
9. ______________
10. ______________
11. ______________
12. ______________
13. ______________
14. ______________
15. ______________
16. ______________
17. ______________
18. ______________
19. ______________
20. ______________
21. ______________
22. ______________
23. ______________
24. ______________
25. ______________
26. ______________
27. ______________
28. ______________
29. ______________
30. ______________
31. ______________
32. ______________
33. ______________
34. ______________

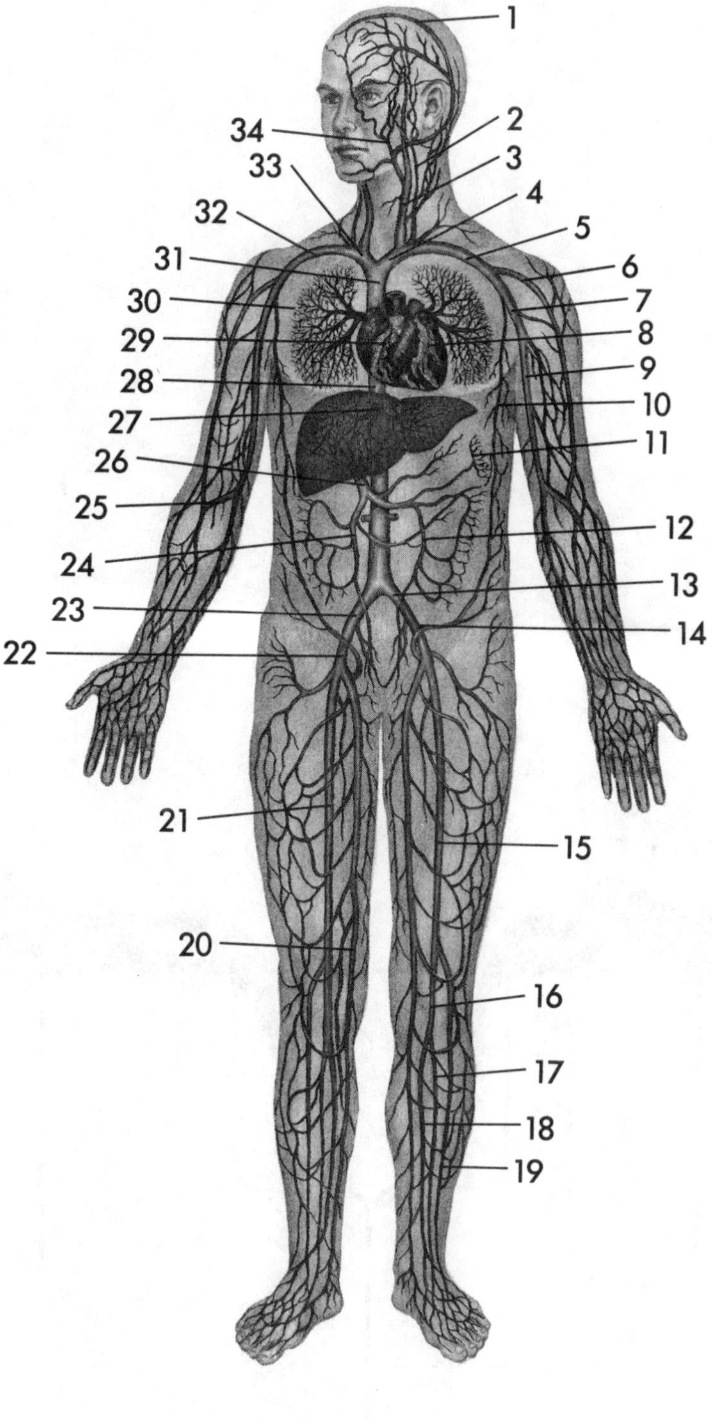

NORMAL ECG DEFLECTIONS

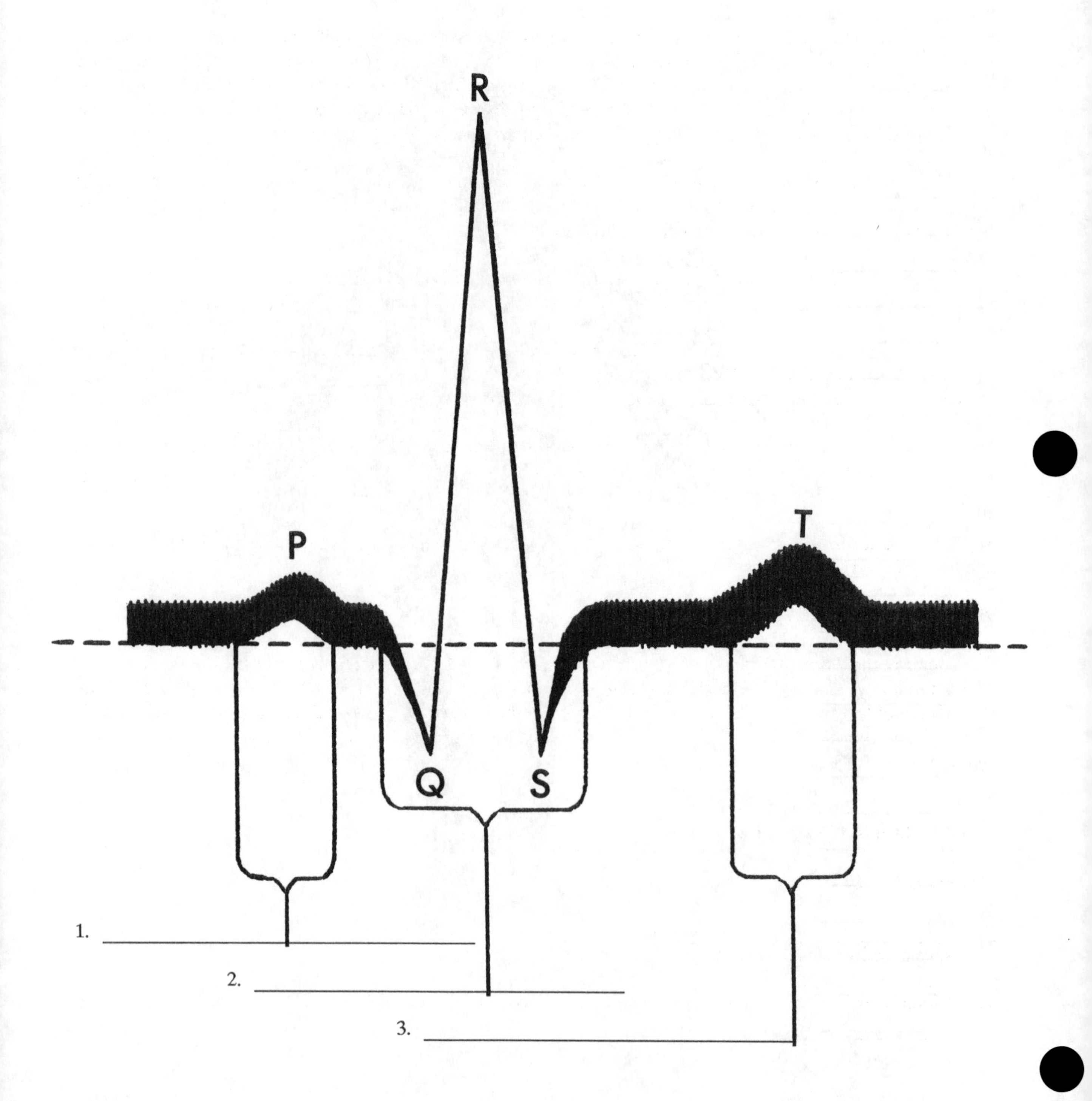

CHAPTER 12

The Lymphatic System and Immunity

The lymphatic system is a system similar to the circulatory system. Lymph, like blood, flows through an elaborate route of vessels. In addition to lymphatic vessels, the lymphatic system consists of lymph nodes, lymph, and the spleen. Unlike the circulatory system, the lymphatic vessels do not form a closed circuit. Lymph flows only once through the vessels before draining into the general blood circulation. This system is a filtering mechanism for microorganisms and serves as a protective device against foreign invaders, such as cancer.

The immune system is the armed forces division of the body. Ready to attack at a moments notice, the immune system defends us against the major enemies of the body: microorganisms, foreign transplanted tissue cells, and our own cells that have turned malignant.

The most numerous cells of the immune system are the lymphocytes. These cells circulate in the body's fluids seeking invading organisms and destroying them with powerful lymphotoxins, lymphokines, or antibodies.

Phagocytes, another large group of immune system cells, assist with the destruction of foreign invaders by a process known as phagocytosis. Neutrophils, monocytes, and connective tissue cells called macrophages use this process to surround unwanted microorganisms, ingest and digest them, rendering them harmless to the body.

Another weapon that the immune system possesses is complement. Normally a group of inactive enzymes present in the blood, complement can be activated to kill invading cells by drilling holes in their cytoplasmic membranes allowing fluid to enter the cell until it bursts.

Your review of this chapter will give you an understanding of how the body defends itself from the daily invasion of destructive substances.

TOPICS FOR REVIEW

Before progressing to Chapter 13 you should familiarize yourself with the functions of the lymphatic system, the immune system, and the major structures that make up these systems. Your review should include knowledge of lymphatic vessels, lymph nodes, lymph, antibodies, complement, and the development of B and T cells. Your study should conclude with an understanding of the differences in humoral and cell-mediated immunity.

THE LYMPHATIC SYSTEM

Fill in the blanks.

1. ____________________ is a specialized fluid formed in the tissue spaces that will be transported by way of specialized vessels to eventually reenter the circulatory system.

2. Blood plasma that has filtered out of capillaries into microscopic spaces between cells is called ______________________.

3. The network of tiny blind-ended tubes distributed in the tissue spaces is called ________________.

4. Lymph eventually empties into two terminal vessels called the __________ and the _____________.

5. The thoracic duct has an enlarged pouchlike structure called the _____________.

6. Lymph is filtered by moving through ________________ which are located in clusters along the pathway of lymphatic vessels.

7. Lymph enters the node through four ___________________ lymph vessels.

8. Lymph exits from the node through a single ______________ lymph vessel.

▸ If you have had difficulty with this section, review pages 252-254. ◂

THYMUS
TONSILS
SPLEEN

Choose the correct response.

(a) Thymus (b) Tonsils (c) Spleen

_____9. Palatine, pharyngeal, and lingual are examples
_____10. Largest lymphoid organ in the body
_____11. Destroys worn out red blood cells
_____12. Located in the mediastinum
_____13. Serves as a reservoir for blood
_____14. T-lymphocytes
_____15. Largest at puberty

▸ If you have had difficulty with this section, review page 256. ◂

THE IMMUNE SYSTEM

Match the term on the left with the proper selection on the right.

_____16. Nonspecific immunity
_____17. Inherited immunity
_____18. Specific immunity
_____19. Acquired immunity
_____20. Immunization

a. Inborn immunity
b. Natural immunity
c. General protection
d. Artificial exposure
e. Memory

IMMUNE SYSTEM MOLECULES

Choose the term that applies to each of the following descriptions. Place the letter for the term in the appropriate answer blank.

a. Antibodies
b. Antigen
c. Allergy
d. Anaphylactic shock
e. Monoclonal
f. Complement fixation
g. Complement
h. Humoral immunity
i. Combining site
j. Macrophage

_____ 21. Hypersensitivity of the immune system to harmless antigens
_____ 22. Life-threatening allergic reaction
_____ 23. Type of very specific antibodies produced from a population of identical cells
_____ 24. Protein compounds normally present in the body
_____ 25. Also known as antibody-mediated immunity
_____ 26. Combines with antibody to produce humoral immunity
_____ 27. Antibody
_____ 28. Process of changing molecule shape slightly to expose binding sites
_____ 29. Phagocyte
_____ 30. Inactive proteins in blood

Circle the one that does not belong.

31. Antibody	Antigen	Protein compound	Combining site
32. Antigen	Invading cells	Foreign protein	Complement
33. Monoclonal	Antibodies	Antigen	Specific
34. Allergy	Complement	Anaphylactic shock	Antigen
35. Monoclonal	14	Complement	Proteins

▸ If you have had difficulty with this section, review pages 255-260. ◂

IMMUNE SYSTEM CELLS

Circle the best answer.

36. The most numerous cells of the immune system are the:

a. Monocytes
b. Eosinophils
c. Neutrophils
d. Lymphocytes
e. Complement

37. Which of the terms listed below occurs third in the immune process?

a. Plasma cells
b. Stem cells
c. Antibodies
d. Activated B cells
e. Immature B cells

38. Which one of the terms listed below occurs last in the immune process?

a. Plasma cells
b. Stem cells
c. Antibodies
d. Activated B cells
e. Immature B cells

39. Moderate exercise has been found to:

a. Decrease white blood cells
b. Increase white blood cells
c. Decrease platelets
d. Decrease red blood cells

40. Which one of the following is part of the cell membrane of B cells?

a. Complement
b. Antigens
c. Antibodies
d. Epitopes
e. None of the above is correct

41. Immature B cells have:

a. Four types of defense mechanisms on their cell membrane
b. Several kinds of defense mechanisms on their cell membrane
c. One specific kind of defense mechanism on their cell membrane
d. No defense mechanisms on their cell membrane

42. Development of an immature B cell depends on the B cell coming in contact with:

a. Complement
b. Antibodies
c. Lymphotoxins
d. Lymphokines
e. Antigens

43. The kind of cell that produces large numbers of antibodies is the:

a. B cell
b. Stem cell
c. T cell
d. Memory cell
e. Plasma cell

44. Just one of these short-lived cells that make antibodies can produce ______________________ of them per second.

a. 20
b. 200
c. 2,000
d. 20,000

45. Which of the following statements is *not* true of memory cells?

a. They produce large numbers of antibodies
b. They are found in lymph nodes
c. They develop into plasma cells
d. They can react with antigens
e. All of the above are true of memory cells

46. T cell development begins in the:

a. Lymph nodes
b. Liver
c. Pancreas
d. Spleen
e. Thymus

47. Human Immunodeficiency Virus (HIV) has its most obvious effects in:

a. B cells
b. Stem cells
c. Plasma cells
d. T cells

48. A milder form of AIDS is known as:

a. Kaposi's sarcoma
b. Pneumocystitis
c. RNA retrovirus
d. AZT
e. ARC

49. B cells function indirectly to produce:

a. Humoral immunity
b. Cell-mediated immunity
c. Lymphotoxins
d. Lymphokines

50. T cells function to produce:

a. Humoral immunity
b. Cell-mediated immunity
c. Antibodies
d. Memory cells

Fill in the blanks.

51. The first stage of development for B cells is called the ______________ ______________.

52. The second stage of B cell development changes an immature B cell into a/an ____________ ____________.

53. ____________ ____________ secrete copious amounts of antibody into the blood—nearly 2,000 antibody molecules for every second they live.

54. T cells are lymphocytes that have undergone their first stage of development in the ____________ ____________.

55. _____________ blocks HIV's ability to reproduce within infected cells.

56. _____________ is a disease caused by a retrovirus that enters the bloodstream and integrates into the DNA of T cell lymphocytes.

57. Some people develop less severe symptoms with AIDS and stay in a stable, but sick condition for years. This less severe stage of AIDS is known as ____________ ____________.

▸ If you have had difficulty with this section, review pages 261-266. ◂

Unscramble the words.

58. NTCMPEOLEM

59. MTMYIUNI

60. OENCLS

61. FNROERTENI

Take the circled letters, unscramble them, and fill in the statement.

What the student was praying for the night before exams.

62.

APPLYING WHAT YOU KNOW

63. Two-year old baby Metcalfe was exposed to chickenpox. He had been a particularly sickly child and so the doctor decided to give him a dose of interferon. What effect was the physician hoping for in baby Metcalfe's case?

64. Sam was a bisexual and an intravenous drug user. He was diagnosed in 1985 as having HTLV-III. What is the current name for this virus?

65. Baby Coyle was born without a thymus gland. Immediate plans were made for a transplant to be performed. In the meantime, baby Coyle was placed in strict isolation. For what reason was he placed in isolation?

DID YOU KNOW?

In 1990 there were 1700 new cases of AIDS in women and 1100 new cases of AIDS in men that were transmitted heterosexually.

LYMPH AND IMMUNITY

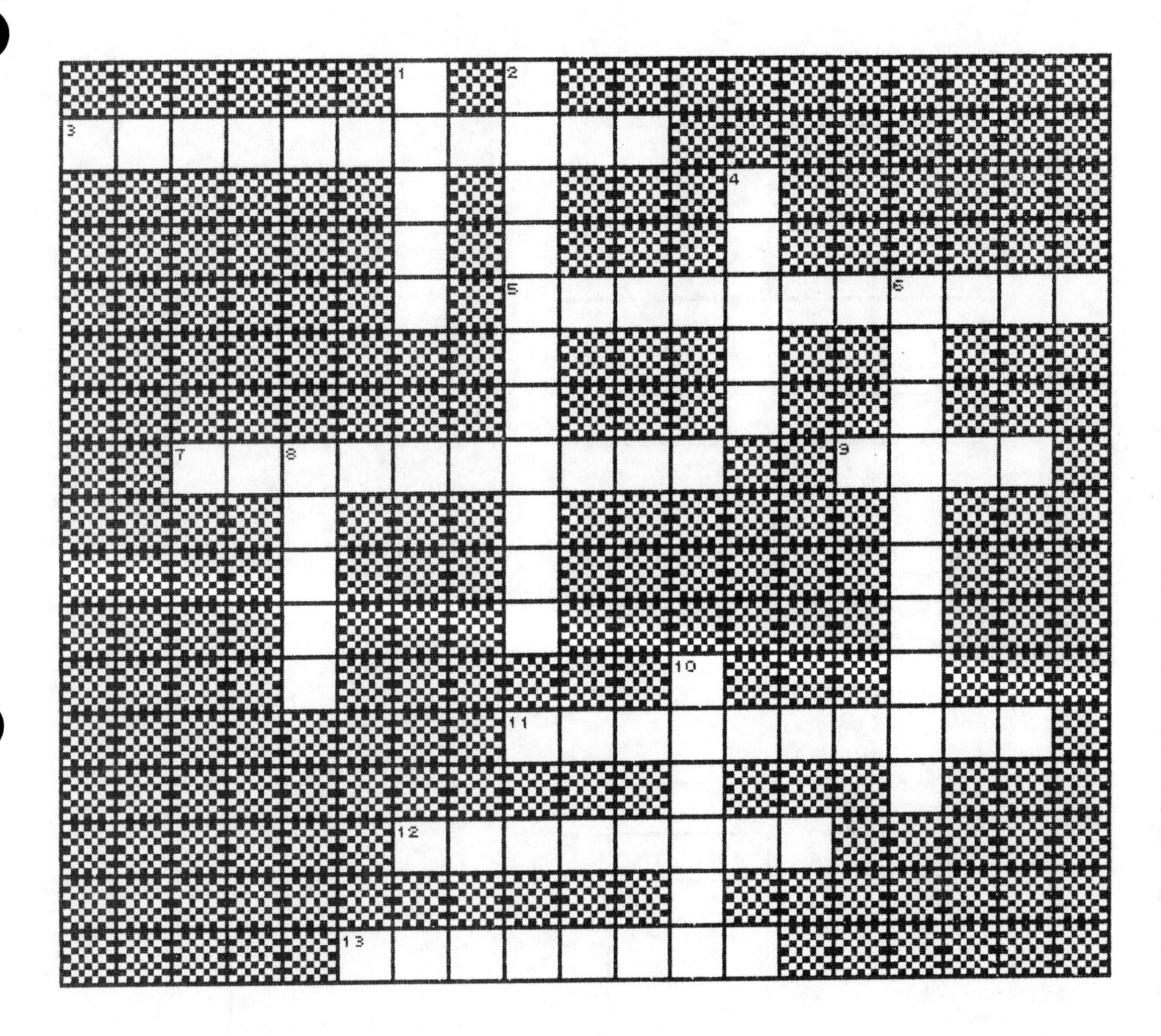

ACROSS

3. Remain in reserve then turn into plasma cells when needed (2 words)
5. Connective tissue cells that are phagocytes
7. Synthetically produced to fight certain diseases
9. Immune deficiency disorder
11. Inactive proteins
12. Lymph enters the node through these lymph vessels
13. Lymph exits the node through this lymph vessel

DOWN

1. Type of lymphocyte (humoral immunity—2 words)
2. Secretes a copious amount of antibodies into the blood (2 words)
4. Family of identical cells descended from one cell
6. Protein compounds normally present in the body
8. Type of lymphocyte (cell-mediated immunity—2 words)
10. Largest lymphoid organ in the body

B CELL DEVELOPMENT

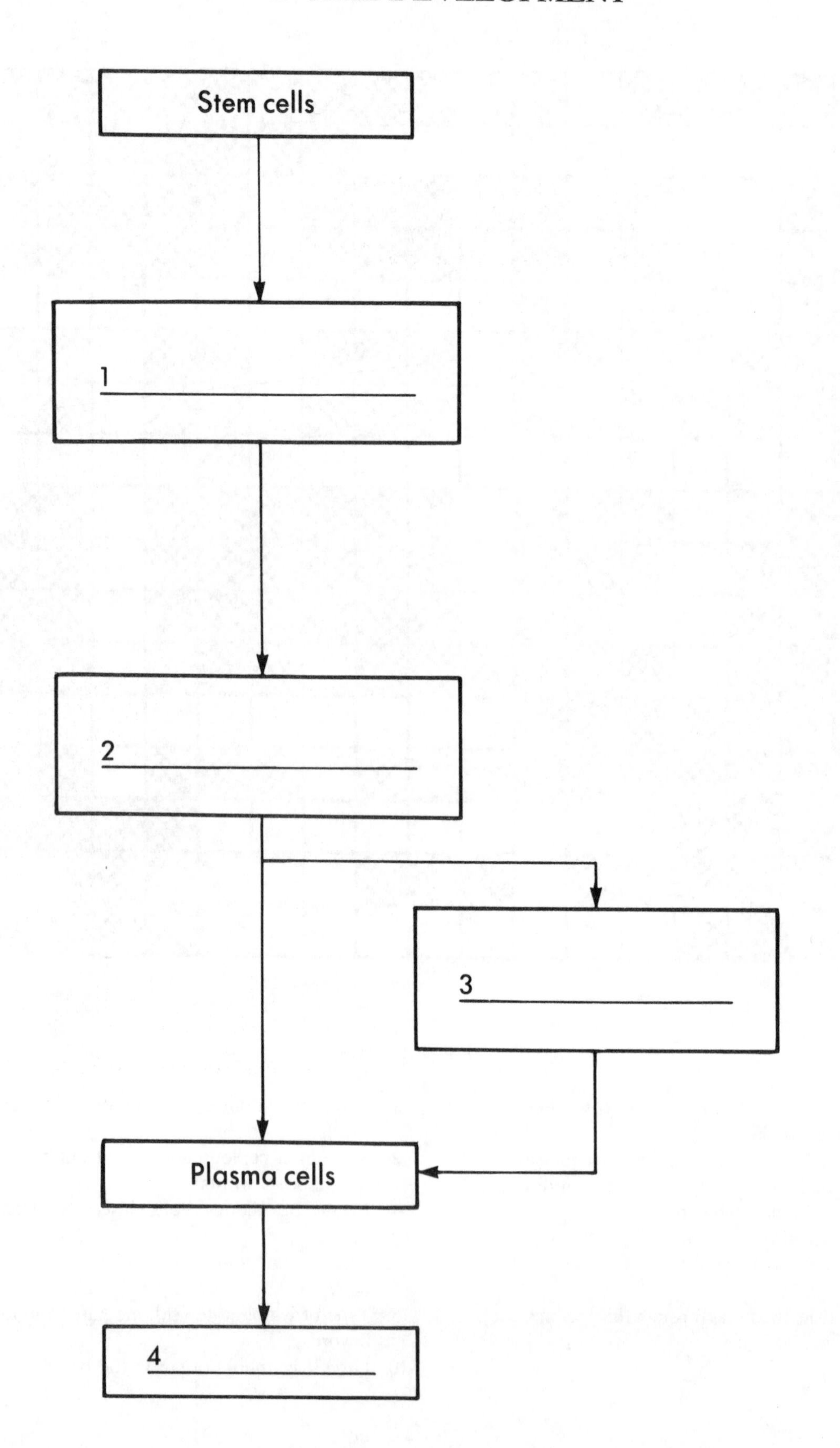

FUNCTION OF SENSITIZED T CELLS

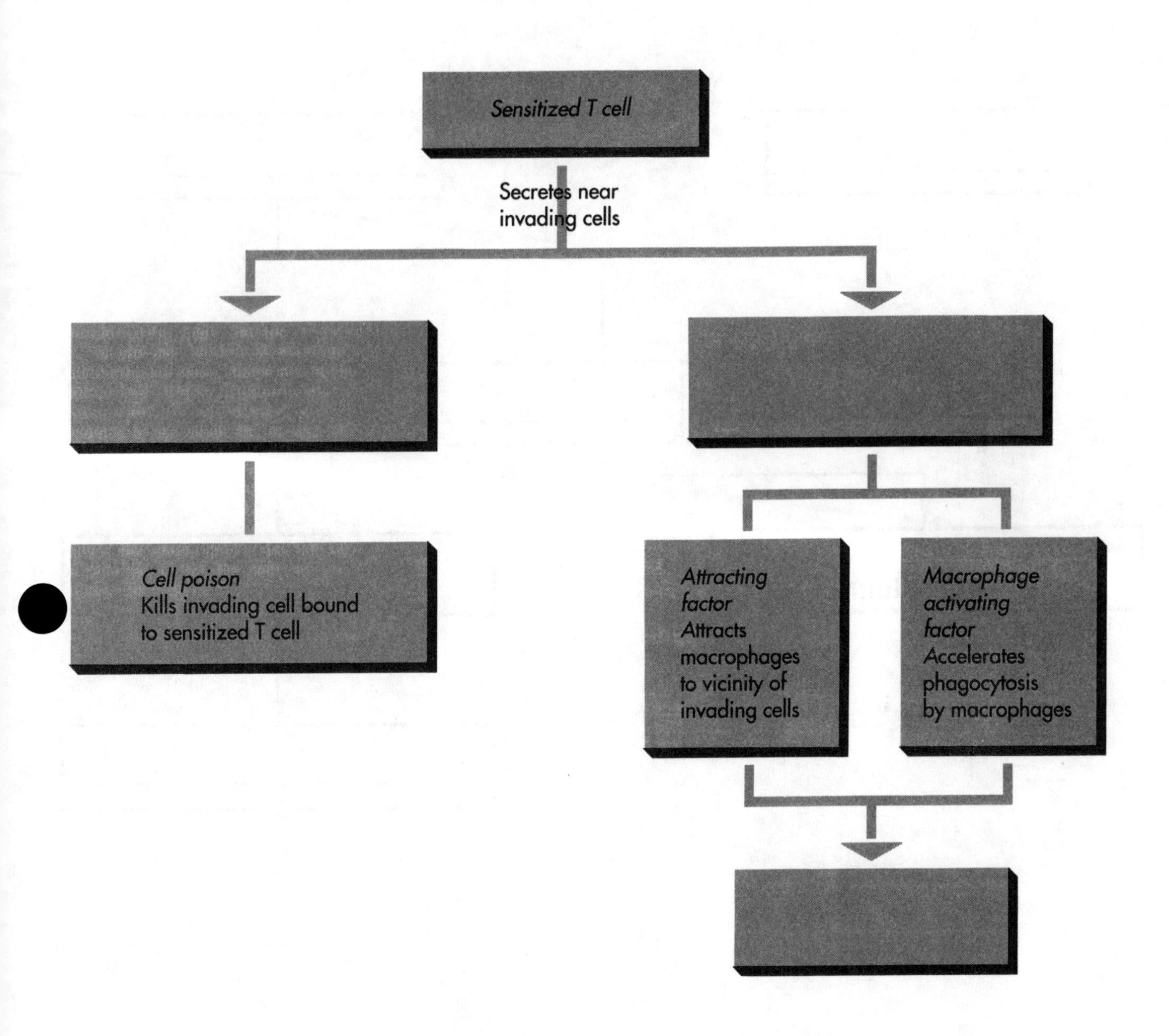

FUNCTION OF ANTIBODIES

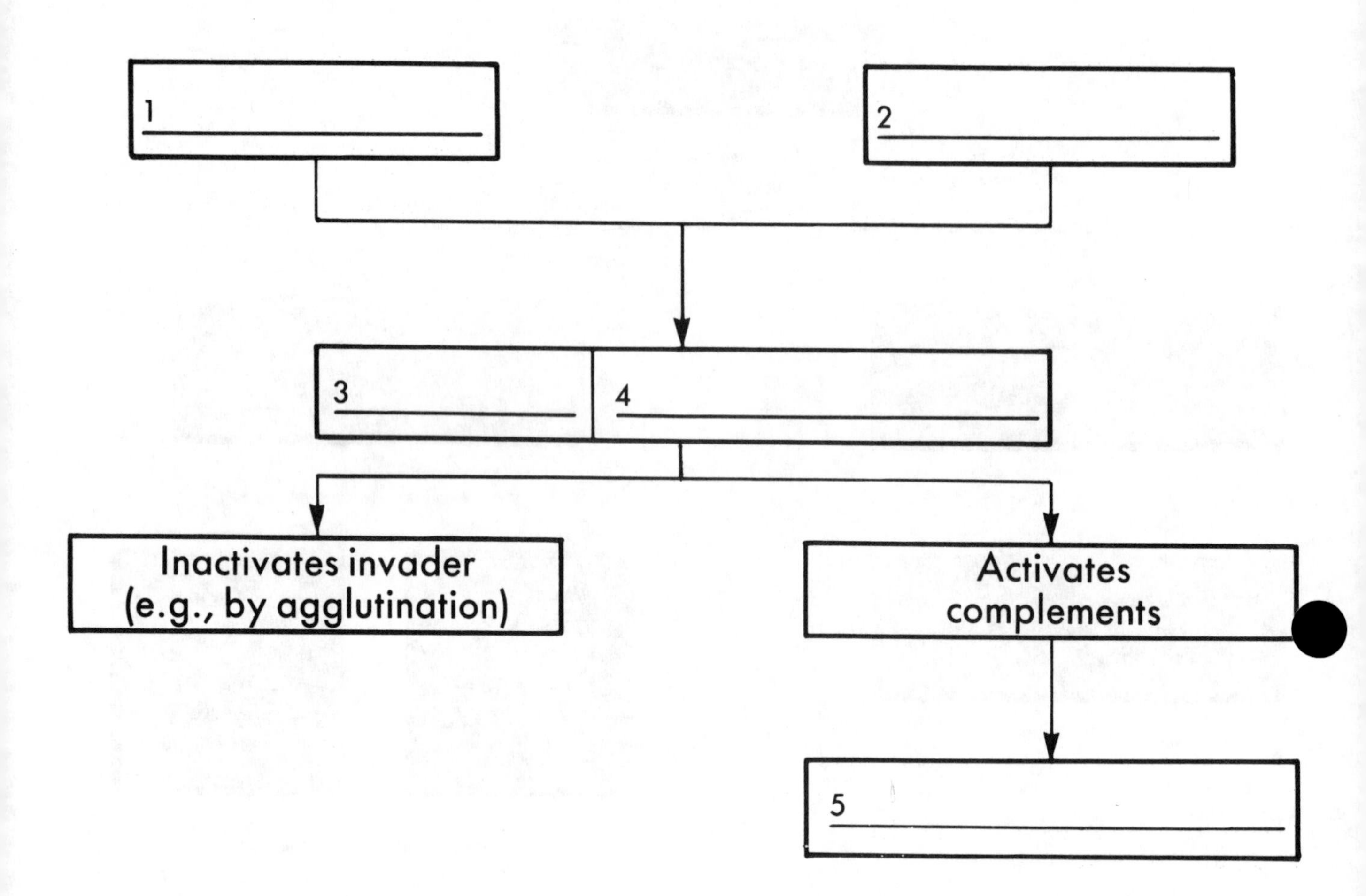

CHAPTER 13

The Respiratory System

As you sit reviewing this system your body needs 16 quarts of air per minute. Walking requires 24 quarts of air, and running requires 50 quarts per minute. The respiratory system provides the air necessary for you to perform your daily activities and eliminates the waste gases from the air that you breath. Take a deep breath, and think of the air as entering some 250 million tiny air sacs similar in appearance to clusters of grapes. These microscopic air sacs expand to let air in and contract to force it out. These tiny sacs, or alveoli, are the functioning units of the respiratory system. They provide the necessary volume of oxygen and eliminate carbon dioxide 24 hours a day.

Air enters either through the mouth or the nasal cavity. It next passes through the pharynx and past the epiglottis, through the glottis and the rest of the larynx. It then continues down the trachea, into the bronchi to the bronchioles, and finally through the alveoli. The reverse occurs for expelled air.

The exchange of gases between air in the lungs and in the blood is known as external respiration. The exchange of gases that occurs between the blood and the cells of the body is known as internal respiration. By constantly supplying adequate oxygen and by removing carbon dioxide as it forms, the respiratory system helps to maintain an environment conducive to maximum cell efficiency.

Your review of this system is necessary to provide you with an understanding of this essential homeostatic mechanism.

TOPICS FOR REVIEW

Before progressing to Chapter 14, you should have an understanding of the structure and function of the organs of the respiratory system. Your review should include knowledge of the mechanisms responsible for both internal and external respiration. Your study should conclude with a knowledge of the volumes of air exchanged in pulmonary ventilation and an understanding of how respiration is regulated.

STRUCTURAL PLAN
RESPIRATORY TRACTS
RESPIRATORY MUCOSA

Match the term with the definition.

a. Diffusion
b. Respiratory membrane
c. Alveoli
d. URI
e. Respiration
f. Respiratory mucosa
g. Upper respiratory tract
h. Lower respiratory tract
i. Cilia
j. Air distributor

_____1. Function of respiratory system
_____2. Pharynx
_____3. Passive transport process responsible for actual exchange of gases
_____4. Assists with the movement of mucus toward the pharynx

_____5. Barrier between the blood in the capillaries and the air in the alveolus
_____6. Lines the tubes of the respiratory tree
_____7. Terminal air sacs
_____8. Trachea
_____9. Head cold
_____10. Homeostatic mechanism

Fill in the blanks.

The organs of the respiratory system are designed to perform two basic functions. They serve as an: (11) __________ __________ and as a (12) __________ __________. In addition to the above, the respiratory system (13) __________, (14) __________, and (15) __________ the air we breath.

Respiratory organs include the (16) __________, (17) __________, (18) __________, (19) __________, (20) __________, and the (21) __________. The respiratory system ends in millions of tiny, thin-walled sacs called (22) __________. (23) __________ of gases takes place in these sacs. Two aspects of the structure of these sacs assist them in the exchange of gases. First, an extremely thin membrane, the (24) __________ __________ allows for easy exchange, and second, the large number of air sacs makes an enormous (25) __________ area.

▸ If you have had difficulty with this section, review pages 272-275. ◂

NOSE
PHARYNX
LARYNX

Circle the word or phrase that does not belong.

26. Nares	Septum	Oropharynx	Conchae
27. Conchae	Frontal	Maxillary	Sphenoidal
28. Oropharynx	Throat	5 inches	Epiglottis
29. Pharyngeal	Adenoids	Uvula	Nasopharynx
30. Middle ear	Tubes	Nasopharynx	Larynx
31. Voice box	Thyroid cartilage	Tonsils	Vocal cords
32. Palatine	Eustachian tube	Tonsils	Oropharynx
33. Pharynx	Epiglottis	Adam's apple	Voice box

Choose the correct response.

(a) Nose (b) Pharynx (c) Larynx

_____34. Warms and humidifies air
_____35. Air and food passes through here
_____36. Sinuses
_____37. Conchae
_____38. Septum
_____39. Tonsils
_____40. Middle ear infections
_____41. Epiglottis

▸ If you have had difficulty with this section, review pages 277-279. ◂

TRACHEA
BRONCHI, BRONCHIOLES AND ALVEOLI
LUNGS AND PLEURA

Fill in the blanks.

42. The windpipe is more properly referred to as the _____________.

43. ___________ keeps the framework of the trachea almost noncollapsible.

44. A lifesaving technique designed to free the trachea of ingested food or foreign objects is the ___________.

45. The first branch or division of the trachea leading to the lungs is the ___________.

46. Each alveolar duct ends in several ____________.

47. The narrow part of each lung, up under the collarbone, is its _____________.

48. The _____________ covers the outer surface of the lungs and lines the inner surface of the rib cage.

49. Inflammation of the lining of the thoracic cavity is _____________.

50. The presence of air in the pleural space on one side of the chest is a ______________.

▸ If you have had difficulty with this section, review pages 279-283. ◂

RESPIRATION

Mark T if the answer is true. If the answer is false, circle the incorrect word(s) and correct the statement.

_________51. Diffusion is the process that moves air into and out of the lungs.
_________52. For inspiration to take place, the diaphragm and other respiratory muscles relax.
_________53. Diffusion is a passive process that results in movement up a concentration gradient.
_________54. The exchange of gases that occurs between blood in tissue capillaries and the body cells is external respiration.
_________55. Many pulmonary volumes can be measured as a person breathes into a spirometer.

__________56. Ordinarily we take about 2 pints of air into our lungs.
__________57. The amount of air normally breathed in and out with each breath is called tidal volume.
__________58. The largest amount of air that one can breathe out in one expiration is called residual volume.
__________59. The inspiratory reserve volume is the amount of air that can be forcibly inhaled after a normal inspiration.

▸ If you have had difficulty with this section, review pages 284-289. ◂

Circle the best answer.

60. The term that means the same thing as breathing is:

a. Gas exchange
b. Respiration
c. Inspiration
d. Expiration
e. Pulmonary ventilation

61. Carbaminohemoglobin is formed when ____________ binds to hemoglobin.

a. Oxygen
b. Amino acids
c. Carbon dioxide
d. Nitrogen
e. None of the above is correct

62. Most of the oxygen transported by the blood is:

a. Dissolved to white blood cells
b. Bound to white blood cells
c. Bound to hemoglobin
d. Bound to carbaminohemoglobin
e. None of the above is correct

63. Which of the following would *not* cause inspiration?

a. Elevation of the ribs
b. Elevation of the diaphragm
c. Contraction of the diaphragm
d. Chest cavity becomes longer from top to bottom

64. A young adult male would have a vital capacity of about __________ ml.

a. 500
b. 1,200
c. 3,300
d. 4,800
e. 6,200

65. The amount of air that can be forcibly exhaled after expiring the tidal volume is known as the:

a. Total lung capacity
b. Vital capacity
c. Inspiratory reserve volume
d. Expiratory reserve volume
e. None of the above is correct

66. Which one of the following is correct?

a. VC = TV - IRV + ERV
b. VC = TV + IRV - ERV
c. VC = TV + IRV x ERV
d. VC = TV + IRV + ERV
e. None of the above is correct

▸ If you have had difficulty with this section, review pages 284-289. ◂

REGULATION OF RESPIRATION
RECEPTORS INFLUENCING RESPIRATION
TYPES OF BREATHING

Match the term on the left with the proper selection on the right.

_____67. Inspiratory center
_____68. Chemoreceptors
_____69. Pulmonary stretch receptors
_____70. Dyspnea
_____71. Respiratory arrest
_____72. Eupnea
_____73. Hypoventilation

a. Difficult breathing
b. Located in carotid bodies
c. Slow and shallow respirations
d. Normal respiratory rate
e. Located in the medulla
f. Failure to resume breathing following a period of apnea
g. Located throughout pulmonary airways and in the alveoli

▸ If you have had difficulty with this section, review pages 290-291. ◂

Unscramble the words.

74. SPUELIRY

75. CRNBOSITHI

76. SESXPTIAI

77. DDNEAOIS

Take the circled letters, unscramble them, and fill in the statement.

What Mona Lisa was to DaVinci.

78.

APPLYING WHAT YOU KNOW

79. Peter is a heavy smoker. Recently he has noticed that when he gets up in the morning, he has a bothersome cough that brings up a large accumulation of mucus. This cough persists for several minutes and then leaves until the next morning. What is an explanation for this problem?

80. Kim was 5 years old and was a mouth breather. She had repeated episodes of tonsillitis and the pediatrician suggested removal of her tonsils and adenoids. He further suggested that the surgery would probably cure her mouth breathing problem. Why is this a possibility?

DID YOU KNOW?

If the alveoli in our lungs were flattened out they would cover a half of a tennis court.

RESPIRATORY SYSTEM

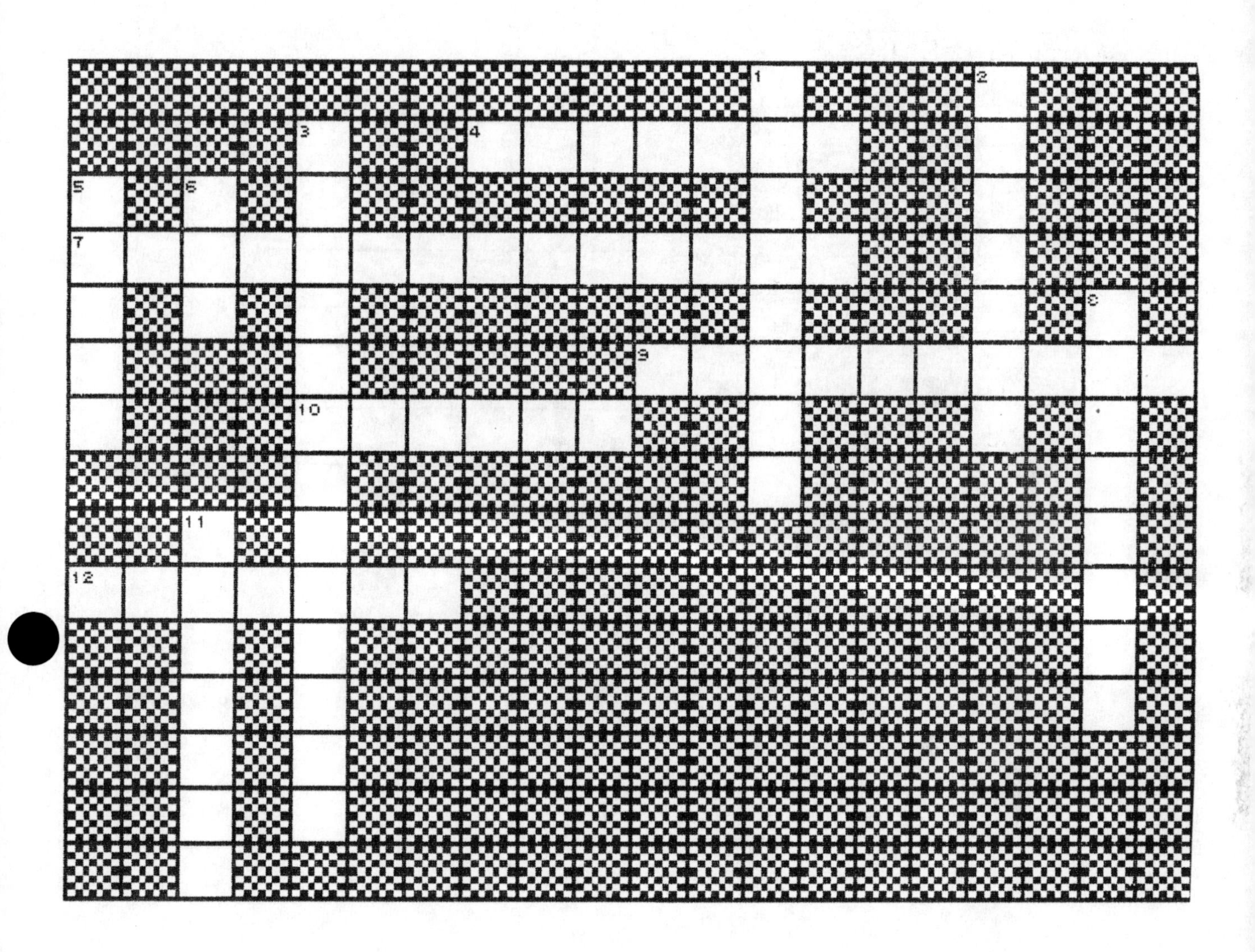

ACROSS

4. Terminal air sacs
7. Sphenoidal (two words)
9. Device used to measure the amount of air exchanged in breathing
10. Voice box
12. Trachea branches into right and left structures

DOWN

1. Inflammation of pleura
2. Windpipe
3. Surgical procedure to remove tonsils
5. Respirations stop
6. Expiratory reserve volume (abbrev.)
8. Doctor who developed lifesaving technique
11. Shelflike structures that protrude into the nasal cavity

SAGITTAL VIEW OF FACE AND NECK

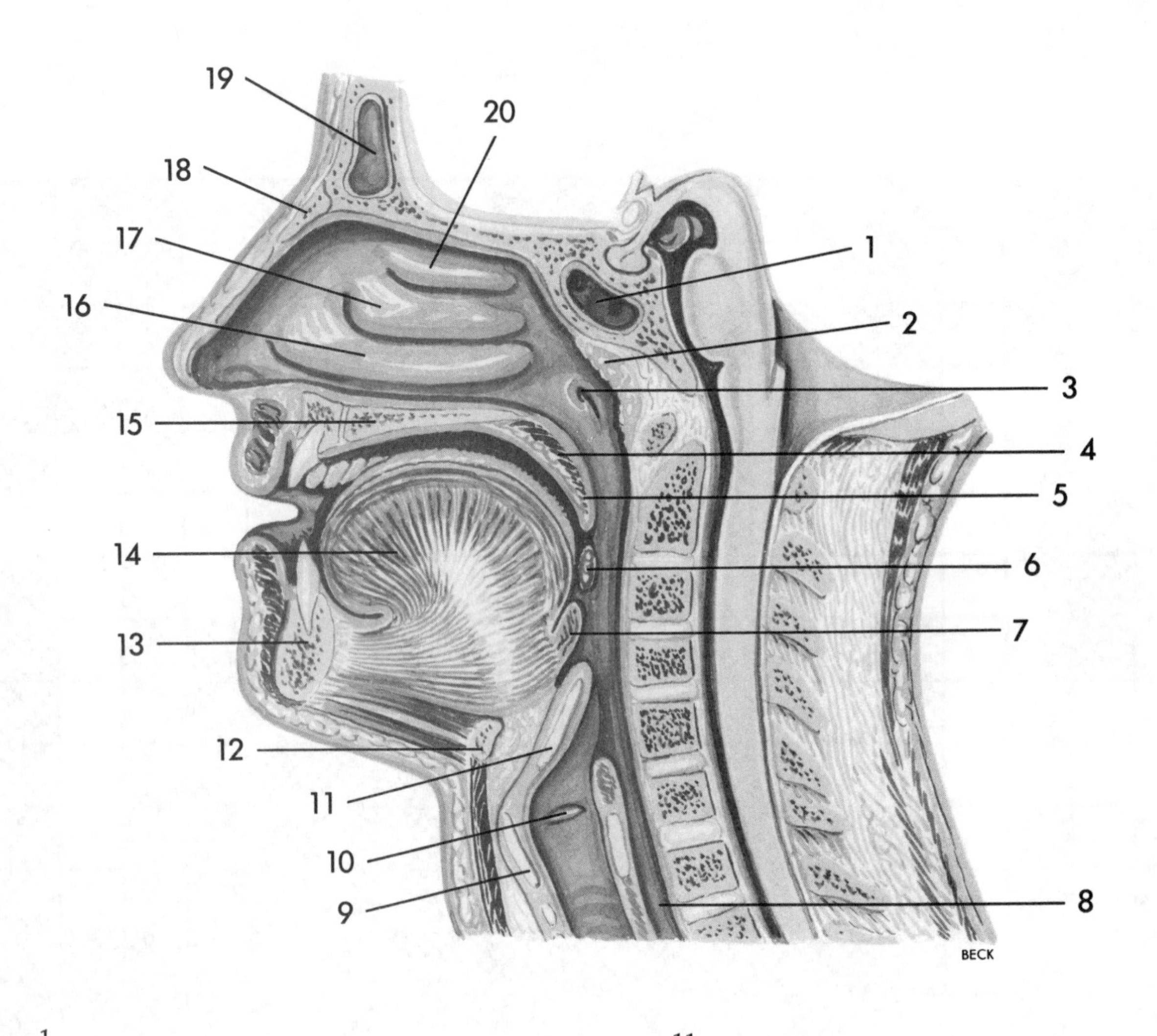

1. ______________________
2. ______________________
3. ______________________
4. ______________________
5. ______________________
6. ______________________
7. ______________________
8. ______________________
9. ______________________
10. ______________________
11. ______________________
12. ______________________
13. ______________________
14. ______________________
15. ______________________
16. ______________________
17. ______________________
18. ______________________
19. ______________________
20. ______________________

RESPIRATORY ORGANS

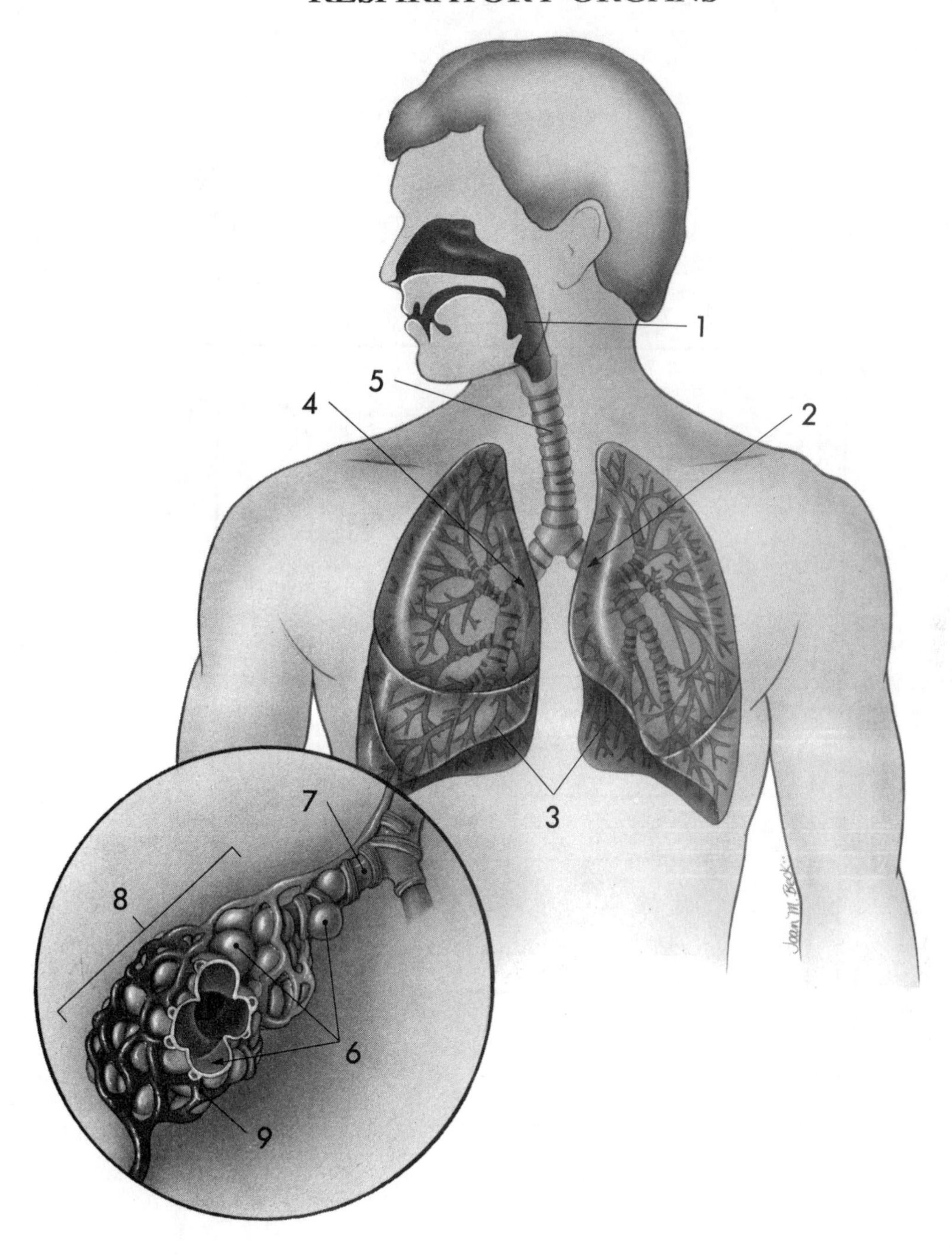

1. ______________________
2. ______________________
3. ______________________
4. ______________________
5. ______________________
6. ______________________
7. ______________________
8. ______________________
9. ______________________

PULMONARY VENTILATION VOLUMES

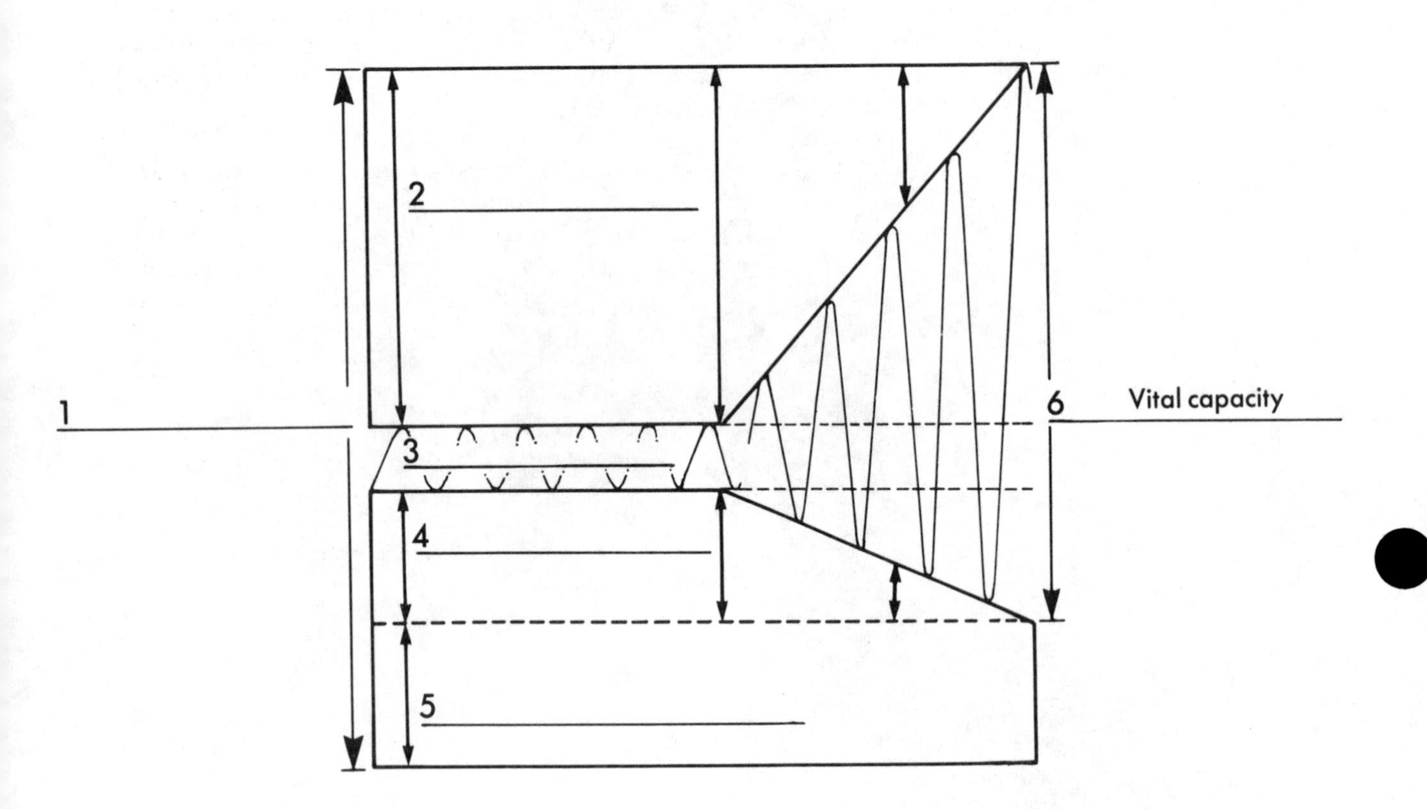

1. ______________________
2. ______________________
3. ______________________
4. ______________________
5. ______________________

CHAPTER 14

The Digestive System

Think of the last meal you ate. The different shapes, sizes, tastes and textures that you so recently enjoyed. Think of those items circulating in your bloodstream in those same original shapes and sizes. Impossible? Of course. And because of this impossibility you will begin to understand and marvel at the close relationship of the digestive system to the circulatory system. It is the digestive system that changes our food, both mechanically and chemically, into a form that is acceptable to the blood and the body.

This change begins the moment you take the very first bite. Digestion starts in the mouth, where food is chewed and mixed with saliva. It then moves down the pharynx and esophagus by peristalsis and enters the stomach. In the stomach it is churned and mixed with gastric juices to become chyme. The chyme goes from the stomach into the duodenum where it is further broken down chemically by intestinal fluids, bile, and pancreatic juice. Those secretions prepare the food for absorption all along the course of the small intestine. Products that are not absorbed pass on through the entire length of the small intestine (duodenum, jejunum, ileum). From there they enter into the cecum of the large intestine, the ascending colon, transverse colon, descending colon, sigmoid colon, into the rectum and out the anus.

Products that are used in the cells undergo absorption. Absorption allows newly processed nutrients to pass through the walls of the digestive tract and into the bloodstream to be distributed to the cells.

Your review of this system will help you understand the mechanical and chemical processes necessary to convert food into energy sources and compounds necessary for survival.

TOPICS FOR REVIEW

Before progressing to Chapter 15, you should review the structure and function of all the organs of digestion. You should have an understanding of the process of digestion, both chemical and mechanical, and of the processes of absorption and metabolism.

WALL OF THE DIGESTIVE SYSTEM

Fill in the blanks.

1. The organs of the digestive system form an irregular-shaped tube called the alimentary canal or ________________ ________________.
2. The churning of food in the stomach is an example of the ______________ breakdown of food.
3. ____________________ breakdown occurs when digestive enzymes act on food as it passes through the digestive tract.
4. Waste material resulting from the digestive process is known as __________________.
5. Foods undergo three kinds of processing in the body: ________________, __________________, and __________________.

6. The serosa of the digestive tube is composed of the ________________ ________________ in the abdominal cavity.

7. The digestive tract extends from the ______________ to the ______________.

8. The inside or hollow space within the alimentary canal is called the ________________.

9. The inside layer of the digestive tract is the ________________.

10. The connective tissue layer that lies beneath the lining of the digestive tract is the ________________.

11. The muscularis contracts and moves food through the gastrointestinal tract by a process known as ________________.

12. The outermost covering of the digestive tube is the ______________.

13. The loops of the digestive tract are anchored to the posterior wall of the abdominal cavity by the ________________.

Select the correct response from the two choices given and insert the letter in the answer blank.

(a) Main organ (b) Accessory organ

_____ 14. Mouth
_____ 15. Parotids
_____ 16. Liver
_____ 17. Stomach
_____ 18. Cecum
_____ 19. Esophagus
_____ 20. Rectum
_____ 21. Pharynx
_____ 22. Appendix
_____ 23. Teeth
_____ 24. Gallbladder
_____ 25. Pancreas

▸ If you have had difficulty with this section, review pages 296-297. ◂

MOUTH
TEETH
SALIVARY GLANDS

Circle the best answer.

26. Which one of the following is *not* a part of the roof of the mouth?

a. Uvula
b. Palatine bones
c. Maxillary bones
d. Soft palate
e. All of the above are part of the roof of the mouth

27. The largest of the papillae on the surface of the tongue are the:

a. Filiform
b. Fungiform
c. Vallate
d. Taste buds

28. The first baby tooth, on an average, appears at:

a. 2 months
b. 1 year
c. 3 months
d. 1 month
e. 6 months

29. The portion of the tooth that is covered with enamel is the:

a. Pulp cavity
b. Neck
c. Root
d. Crown
e. None of the above is correct

30. The wall of the pulp cavity is surrounded by:

a. Enamel
b. Dentin
c. Cementum
d. Connective tissue
e. Blood and lymphatic vessels

31. Which of the following teeth is missing from the deciduous arch?

a. Central incisor
b. Canine
c. Second premolar
d. First molar
e. Second molar

32. The permanent central incisor erupts between the ages of ____________.

a. 9-13
b. 5-6
c. 7-10
d. 7-8
e. None of the above is correct

33. The third molar appears between the ages of __________________.

a. 10-14
b. 5-8
c. 11-16
d. 17-24
e. None of the above is correct

34. Which one of the following will *not* significantly reduce caries?

a. Reducing the amount of refined sugar in the diet
b. Fluoride in the water supply
c. Regular and thorough brushing
d. Eating a carrot or stick of celery instead of brushing

35. The ducts of the ________________ glands open into the floor of the mouth.

a. Sublingual
b. Submandibular
c. Parotid
d. Carotid

36. The volume of saliva secreted per day is about:

a. One half pint
b. One pint
c. One liter
d. One gallon

37. Mumps are an infection of the:

a. Parotid gland
b. Sublingual gland
c. Submandibular gland
d. Tonsils

38. Incisors are used during mastication to:

a. Cut
b. Piece
c. Tear
d. Grind

39. Another name for the third molar is:

a. Central incisor
b. Wisdom tooth
c. Canine
d. Lateral incisor

40. After food has been chewed, it is formed into a small rounded mass called a:

a. Moat
b. Chyme
c. Bolus
d. Protease

▸ If you have had difficulty with this section, review pages 298-303. ◂

PHARYNX
ESOPHAGUS
STOMACH

Fill in the blanks.

The (41) ______________ is a tubelike structure that functions as part of both respiratory and digestive systems. It connects the mouth with the (42) __________________. The esophagus serves as a passageway for movement of food from the pharynx to the (43) ________________. Food enters the stomach by passing through the muscular (44) ______________ ________________ at the end of the esophagus. Contraction of the stomach mixes the food thoroughly with the gastric juices and breaks it down into a semisolid mixture called (45) ___________________.

The three divisions of the stomach are the (46) _______________, (47) _______________, and (48) ________________.

Food is held in the stomach by the (49) ______________ ______________ muscle long enough for partial digestion to occur. After food has been in the stomach for approximately 3 hours, the chyme will enter the (50) ______________.

Match the term with the correct definition.

a. Esophagus
b. Chyme
c. Peristalsis
d. Rugae
e. Ulcer
f. Greater curvature
g. Emesis
h. Tagamet
i. Duodenum
j. Lesser curvature

_____ 51. Stomach folds
_____ 52. Upper right border of stomach
_____ 53. Total emptying of stomach contents back through the cardiac sphincter, up the esophagus, and out of the mouth
_____ 54. 10-inch passageway
_____ 55. Drug used to treat ulcers
_____ 56. Semisolid mixture of stomach contents
_____ 57. Muscle contractions of the digestive system
_____ 58. Open wound in digestive system that is acted on by acidic gastric juice
_____ 59. First part of small intestine
_____ 60. Lower left border of stomach

▸ If you have had difficulty with this section, review pages 303-305. ◂

SMALL INTESTINE
LIVER AND GALLBLADDER
PANCREAS

Circle the best answer.

61. Which one is *not* part of the small intestine?

a. Jejunum
b. Ileum
c. Cecum
d. Duodenum

62. Which one of the following structures does *not* increase the surface area of the intestine for absorption?

a. Plicae
b. Rugae
c. Villi
d. Brush border

63. The union of the cystic duct and hepatic duct form the:

a. Common bile duct
b. Major duodenal papilla
c. Minor duodenal papilla
d. Pancreatic duct

64. Obstruction of the _________________ will lead to jaundice.

a. Hepatic duct
b. Pancreatic duct
c. Cystic duct
d. None of the above

65. Each villus in the intestine contains a lymphatic vessel or _________________ that serves to absorb lipid or fat materials from the chyme:

a. Plica
b. Lacteal
c. Villa
d. Microvilli

66. The middle third of the duodenum contains the:

a. Islets
b. Fundus
c. Body
d. Rugae
e. Major duodenal papilla

67. Excessive secretion of acid or _______________ is an important factor in the formation of ulcers.

a. Hypoacidity
b. Emesis
c. Cholecystokinin
d. Hyperacidity

68. The liver is an:

a. Enzyme
b. Endocrine organ
c. Endocrine gland
d. Exocrine gland

69. Fats in chyme stimulate the secretion of the hormone:

a. Lipase
b. Cholecystokinin
c. Protease
d. Amylase

70. The largest gland in the body is the:

a. Pituitary
b. Thyroid
c. Liver
d. Thymus

▸ If you have had difficulty with this section, review pages 305-308. ◂

LARGE INTESTINE
APPENDIX
PERITONEUM

If the statement is true, mark T next to the answer. If the statement is false, circle the incorrect word(s) and write the correct term in the blank next to the statement.

_________71. Bacteria in the large intestine are responsible for the synthesis of vitamin E needed for normal blood clotting.
_________72. Villi in the large intestine absorb salts and water.
_________73. If waste products pass rapidly through the large intestine, constipation results.
_________74. The ileocecal valve opens into the sigmoid colon.
_________75. The splenic flexure is the bend between the ascending colon and the transverse colon.

_________76. The splenic colon is the S-shaped segment that terminates in the rectum.
_________77. The appendix serves no important digestive function in humans.
_________78. For patients with suspected appendicitis, a physician will often evaluate the appendix by a digital rectal examination.
_________79. The visceral layer of the peritoneum lines the abdominal cavity.
_________80. The greater omentum is shaped like a fan and serves to anchor the small intestine to the posterior abdominal wall.

▸ If you have had difficulty with this section, review pages 310-312. ◂

DIGESTION
ABSORPTION
METABOLISM

Circle the best answer.

81. Which one of the following substances does *not* contain any enzymes?

   a. Saliva
   b. Bile
   c. Gastric juice
   d. Pancreatic juice
   e. Intestinal juice

82. Which one of the following is a simple sugar?

   a. Maltose
   b. Sucrose
   c. Lactose
   d. Glucose
   e. Starch

83. Cane sugar is the same as:

   a. Maltose
   b. Lactose
   c. Sucrose
   d. Glucose
   e. None of the above is correct

84. Most of the digestion of carbohydrates takes place in the:

   a. Mouth
   b. Stomach
   c. Small intestine
   d. Large intestine

85. Fats are broken down into:

   a. Amino acids
   b. Simple sugars
   c. Fatty acids
   d. Disaccharides

▸ If you have had difficulty with this section, review page 313. ◂

86. *Fill in the blank areas on the chart below.*

CHEMICAL DIGESTION

DIGESTIVE JUICES AND ENZYMES	SUBSTANCE DIGESTED (OR HYDROLYZED)	RESULTING PRODUCT
SALIVA		
1. Amylase	1.	1. Maltose
GASTRIC JUICE		
2. Protease (pepsin) plus hydrochloric acid	2. Proteins	2.
PANCREATIC JUICE		
3. Protease (trypsin)	3. Amylase Proteins (intact or partially digested)	3.
4. Lipase	4.	4. Fatty acids and glycerol
INTESTINAL JUICE		
5. Peptidases	5.	5. Amino acids
6.	6. Sucrose	6. Glucose and fructose
7. Lactase	7.	7. Glucose and galactose (simple sugars)
8. Maltase	8. Maltose	8.

▸ If you have had difficulty with this section, review page 314. ◂

APPLYING WHAT YOU KNOW

87. Mr. Gabriel was a successful businessman, but he worked too hard and was always under great stress. His doctor cautioned him that if he did not alter his style of living he would be subject to hyperacidity. What could be the resulting condition of hyperacidity?

88. Baby Shearer has been regurgitating his bottle feeding at every meal. The milk is curdled, but does not appear to be digested. He has become dehydrated, and so his mother is taking him to the pediatrician. What is a possible diagnosis from your textbook reading?

89. Mr. Josten has gained a great deal of weight suddenly. He also noticed that he was sluggish and always tired. What test might his physician order for him and for what reason?

90. **WORD FIND**

Can you find the terms listed below in the box of letters? Words may be spelled top to bottom, bottom to top, right to left, left to right, or diagonally.

```
W J A D L B T D N M X H R G S P R U B E
F B Q I M Z K N X O E C H B E T S V P C
H M D L V Y N J R I I Z D W E Y C U B I
S S E P F K O S W U D T E M G P W L T D
B A T A E K I E L M B N A J F Y D A F N
D E E P C W T D G F I T E C L J B I T U
D R O I E H S E P Z Y Q R P I L V M N A
I C L L S E E N Q W T P A A P T G J B J
A N C L C U G T W H C F L F E A S O H P
R A A A H V I I E R Q K U X Y H O A S E
R P V E N F D N Y B U N B N Q E J N M R
H Z I U K W V J D S D E W Q T M N E M I
E E T N R N O U R U I A S O C U M Y E S
A F Y Z I M O R S F R D W D C L J U S T
G O J C J D S R C X C H C M I S Z G E A
I W U N E N O I T P R O S B A I C W N L
R D U N B S T O M A C H N F L F X Y T S
J L U H L I A M Q F Q F G T A Y B Z E I
J M Y L E Q A Y F J Z P J W Q C O J R S
Y M E T A B O L I S M M F T M Y M Q Y S
```

LIST OF WORDS

HEARTBURN
ABSORPTION
PERISTALSIS
MASTICATION
CAVITY
EMULSIFY
FUNDUS
MESENTERY

FECES
METABOLISM
PAPILLAE
CROWN
STOMACH
JAUNDICE
DIARRHEA

DIGESTION
MUCOSA
UVULA
DENTIN
PANCREAS
DUODENUM
APPENDIX

DID YOU KNOW?

The liver performs over 500 functions and produces over 1000 enzymes to handle the chemical conversions necessary for survival.

DIGESTIVE SYSTEM

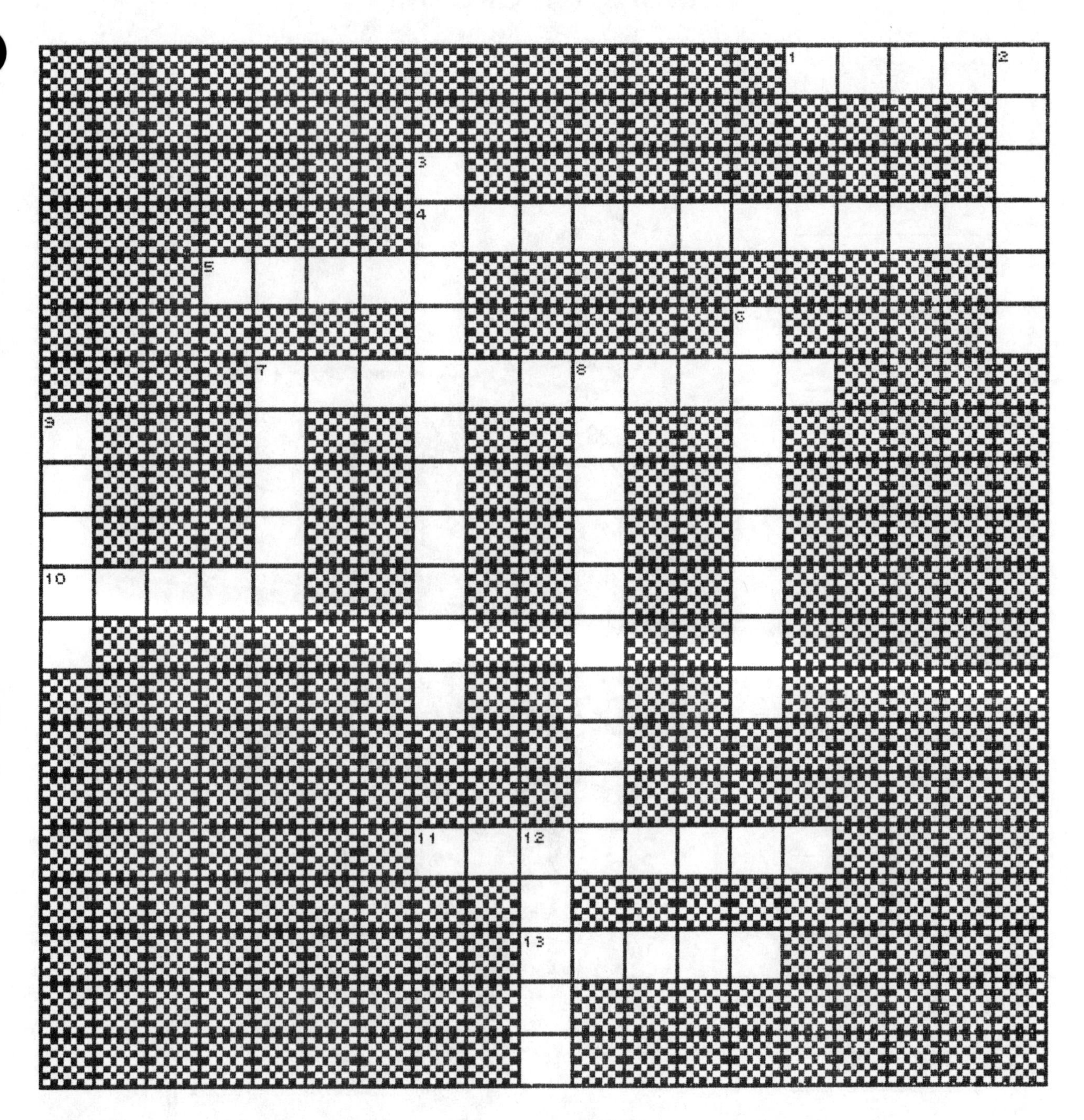

ACROSS

1. Stomach folds
4. Inflammation of the appendix
5. Waste product of digestion
7. Movement of food through digestive tract
10. Prevents food from entering nasal cavities
11. Yellowish skin discoloration
13. Semisolid mixture

DOWN

2. Vomitus
3. Process of chewing
6. Fluid stools
7. Intestinal folds
8. Digested food moves from intestine to blood
9. Rounded mass of food
12. Open wound in digestive area acted on by acid juices

DIGESTIVE ORGANS

1. ____________________
2. ____________________
3. ____________________
4. ____________________
5. ____________________
6. ____________________
7. ____________________
8. ____________________
9. ____________________
10. ____________________
11. ____________________
12. ____________________
13. ____________________
14. ____________________
15. ____________________
16. ____________________
17. ____________________
18. ____________________
19. ____________________
20. ____________________
21. ____________________
22. ____________________
23. ____________________
24. ____________________
25. ____________________
26. ____________________
27. ____________________
28. ____________________
29. ____________________
30. ____________________

TOOTH

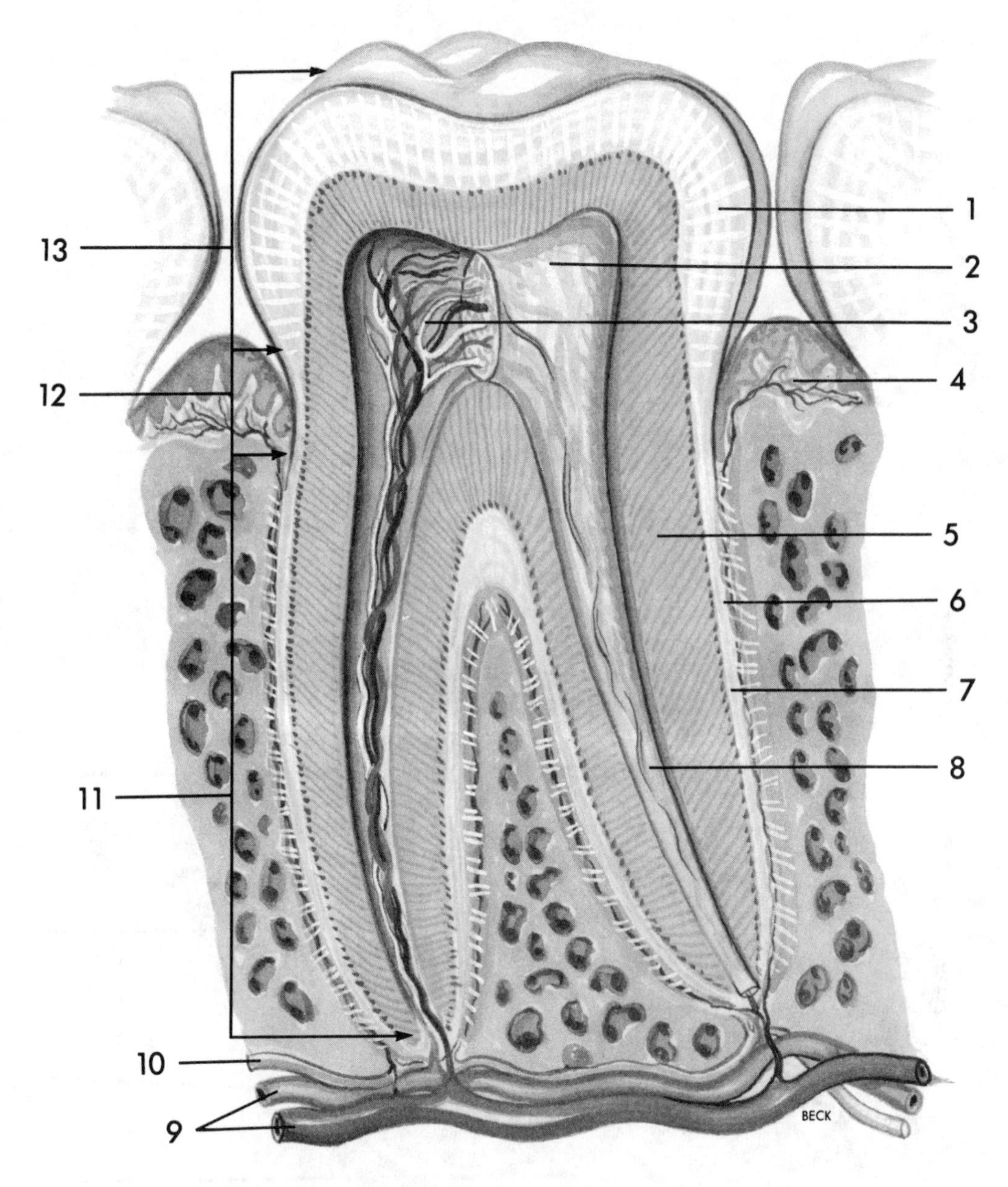

1. ______________________

2. ______________________

3. ______________________

4. ______________________

5. ______________________

6. ______________________

7. ______________________

8. ______________________

9. ______________________

10. ______________________

11. ______________________

12. ______________________

13. ______________________

THE SALIVARY GLANDS

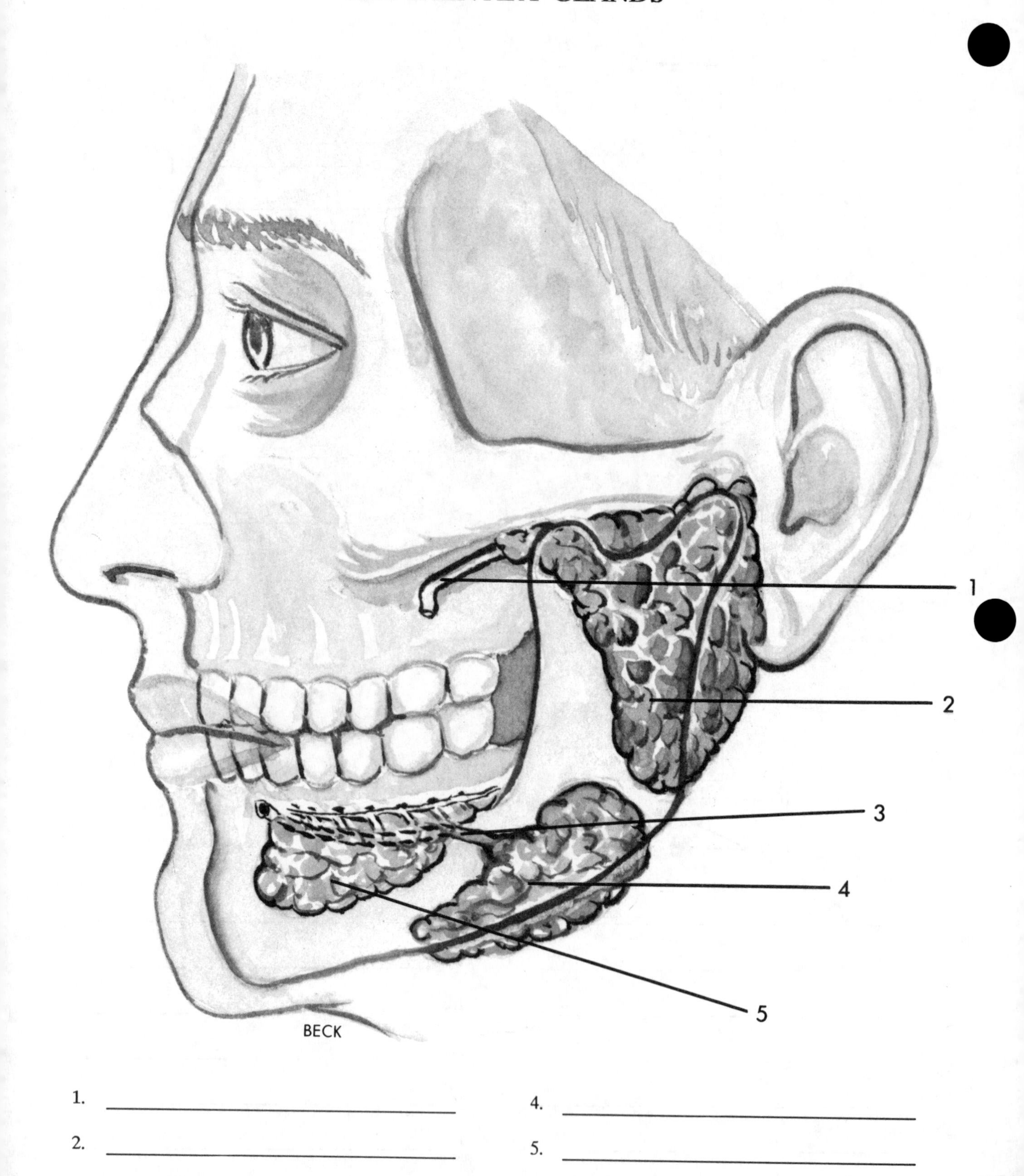

1. ______________________

2. ______________________

3. ______________________

4. ______________________

5. ______________________

STOMACH

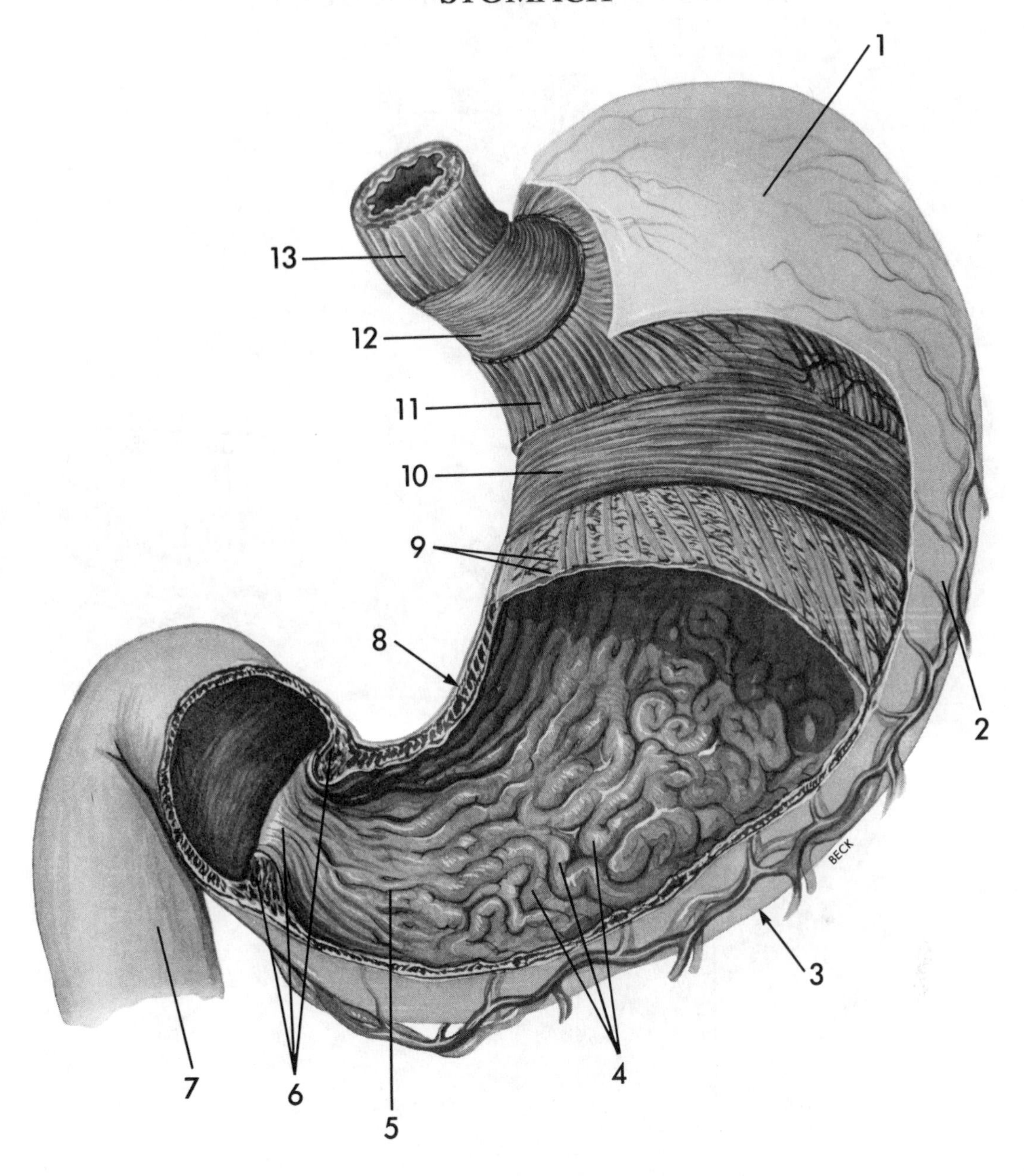

1. ______________________
2. ______________________
3. ______________________
4. ______________________
5. ______________________
6. ______________________
7. ______________________
8. ______________________
9. ______________________
10. ______________________
11. ______________________
12. ______________________
13. ______________________

GALLBLADDER AND BILE DUCTS

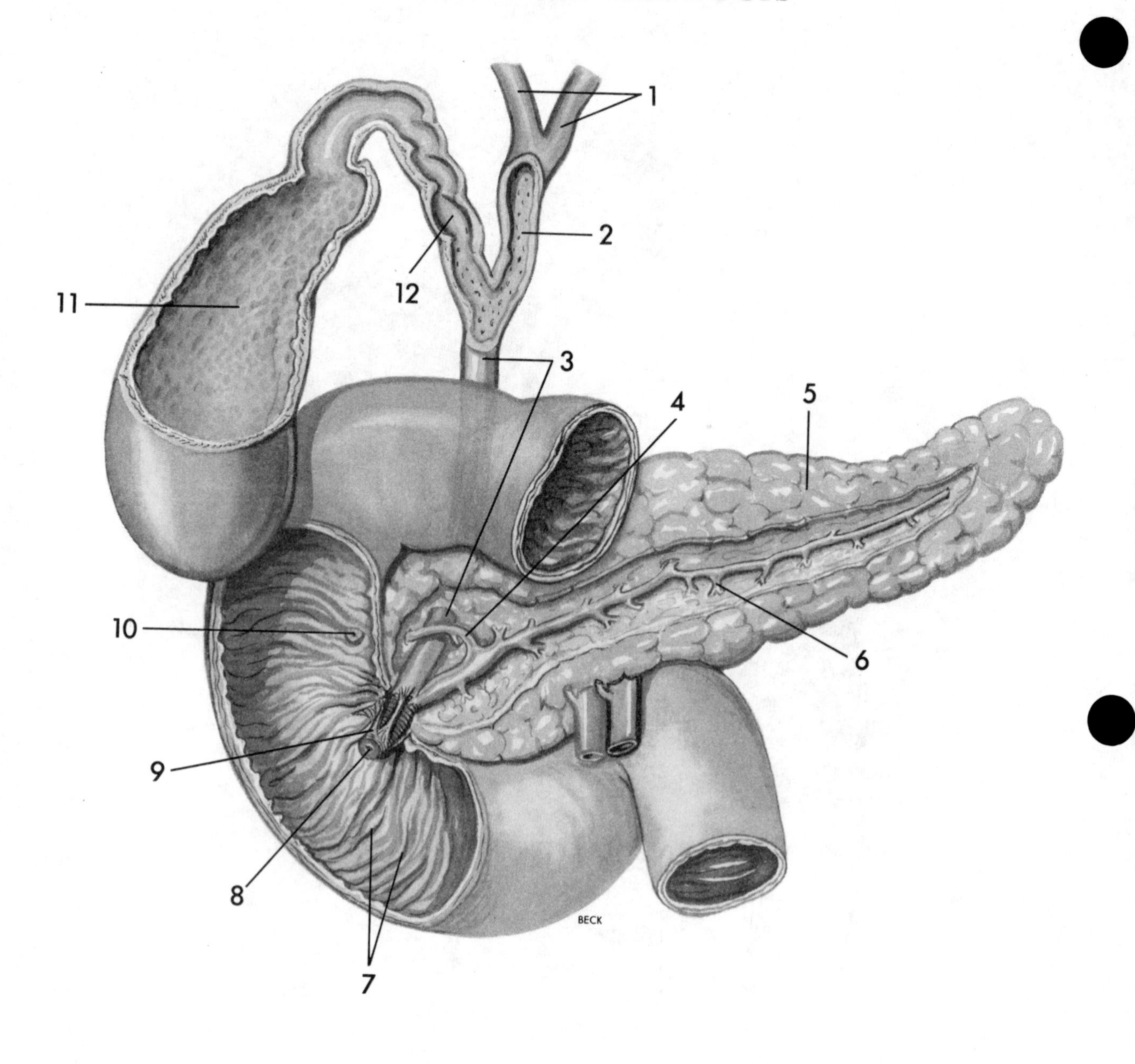

1. ______________________
2. ______________________
3. ______________________
4. ______________________
5. ______________________
6. ______________________
7. ______________________
8. ______________________
9. ______________________
10. ______________________
11. ______________________
12. ______________________

THE SMALL INTESTINE

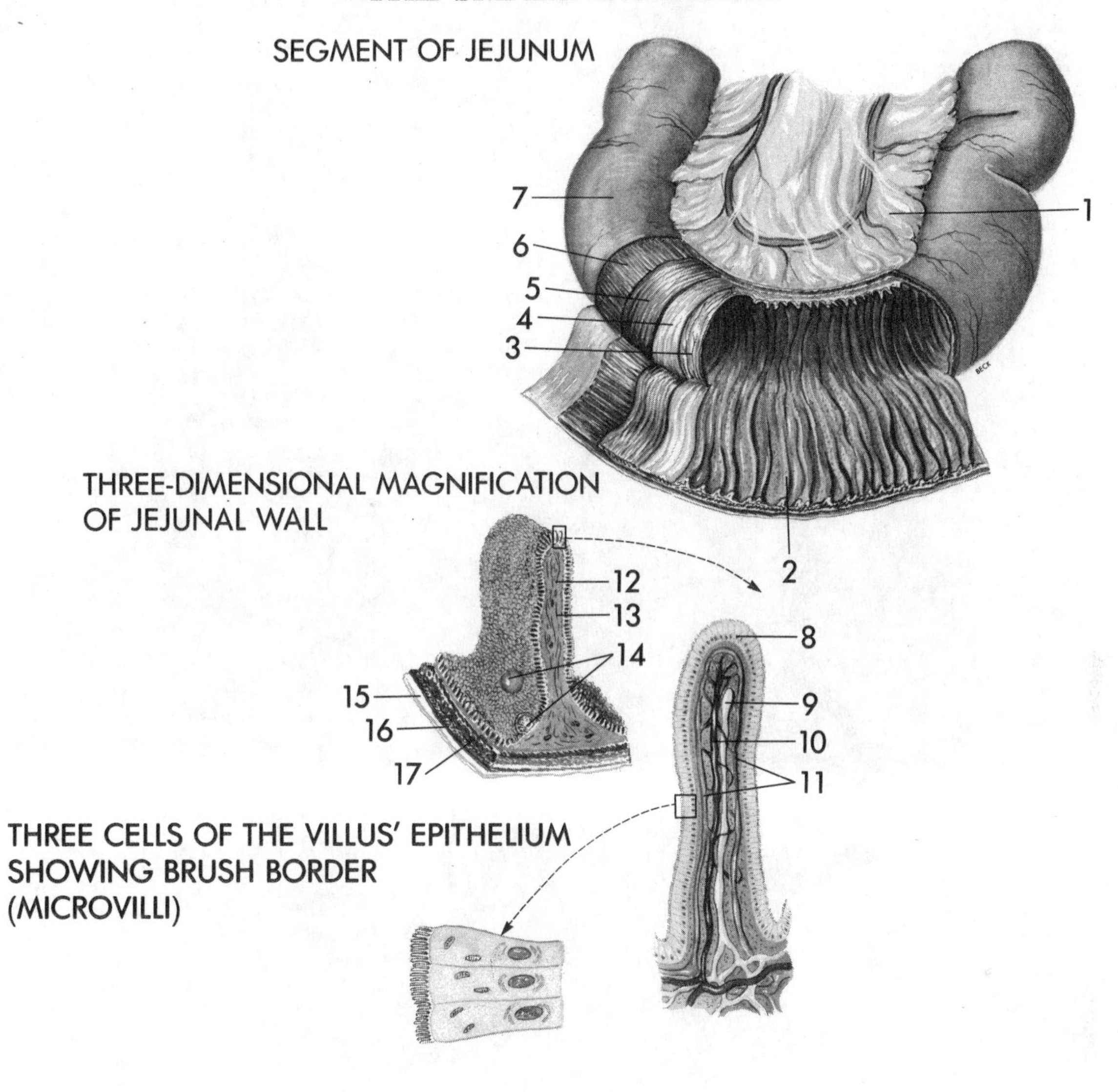

1. ______________________________
2. ______________________________
3. ______________________________
4. ______________________________
5. ______________________________
6. ______________________________
7. ______________________________
8. ______________________________
9. ______________________________
10. ______________________________
11. ______________________________
12. ______________________________
13. ______________________________
14. ______________________________
15. ______________________________
16. ______________________________
17. ______________________________

THE LARGE INTESTINE

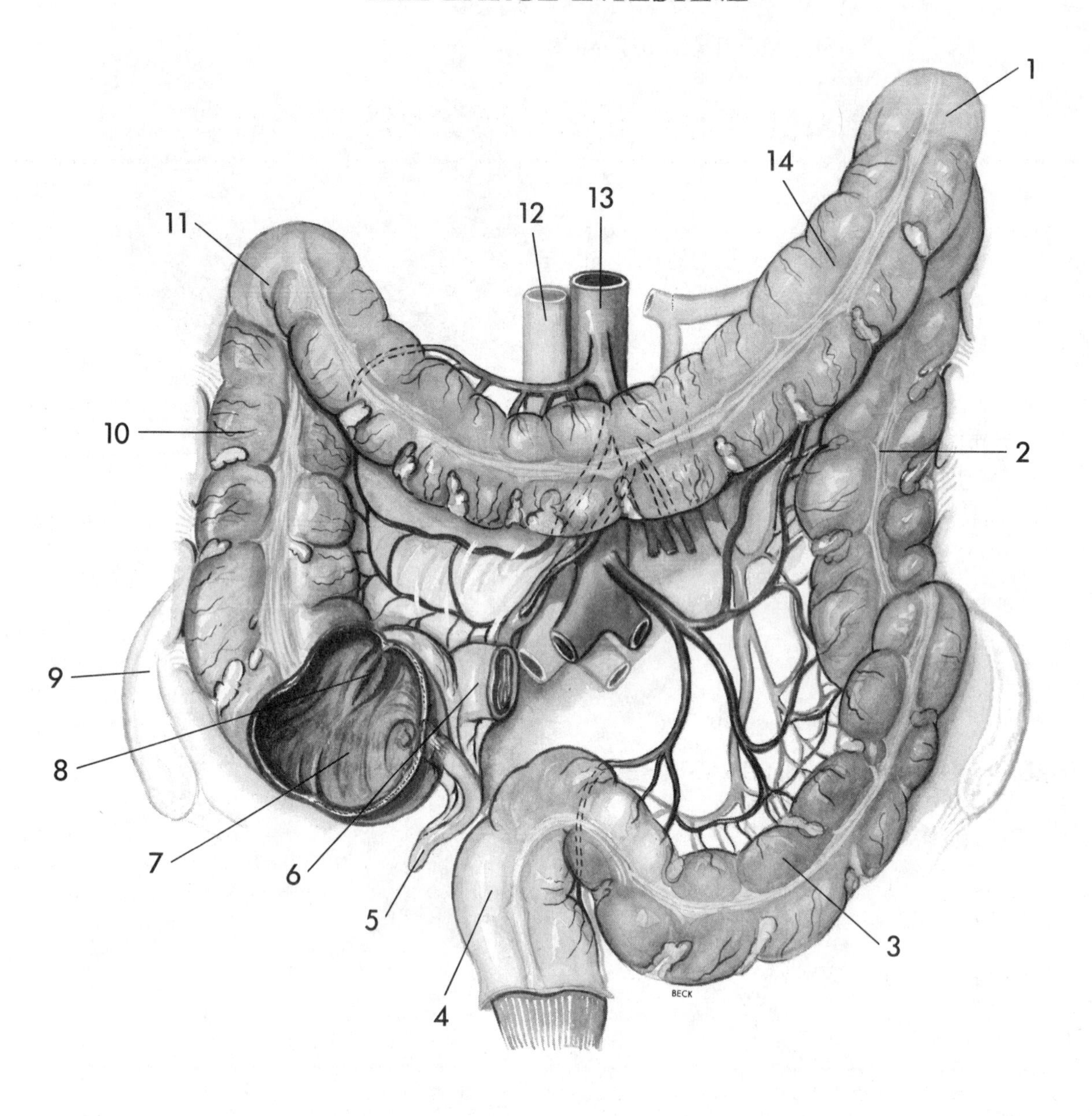

1. ______________________
2. ______________________
3. ______________________
4. ______________________
5. ______________________
6. ______________________
7. ______________________
8. ______________________
9. ______________________
10. ______________________
11. ______________________
12. ______________________
13. ______________________
14. ______________________

CHAPTER 15

Nutrition and Metabolism

Most of us love to eat, but do the foods we enjoy provide us with the basic food types necessary for good nutrition? The body, a finely tuned machine, requires a balance of carbohydrates, fats, proteins, vitamins, and minerals to function properly. These nutrients must be digested, absorbed, and circulated to cells constantly to accommodate the numerous activities that occur throughout the body. The use the body makes of foods once these processes are completed is called "metabolism."

The liver plays a major role in the metabolism of food. It helps maintain a normal blood glucose level, removes toxins from the blood, processes blood immediately after it leaves the gastrointestinal tract, and initiates the first steps of protein and fat metabolism.

This chapter also discusses basal metabolic rate (BMR). The BMR is the rate at which food is catabolized under basal conditions. This test and the protein-bound iodine (PBI) are indirect measures of thyroid gland functioning. The total metabolic rate (TMR) is the amount of energy, expressed in calories, used by the body each day.

Finally, maintaining a constant body temperature is a function of the hypothalamus and a challenge for the metabolic factors of the body. Review of this chapter is necessary to provide you with an understanding of the "fuel" or nutrition necessary to maintain your complex homeostatic machine—the body.

TOPICS FOR REVIEW

Before progressing to Chapter 16, you should be able to define and contrast catabolism and anabolism. Your review should include the metabolic roles of carbohydrates, fats, proteins, vitamins, and minerals. Your study should conclude with an understanding of the basal metabolic rate and physiological mechanisms that regulate body temperature.

THE ROLE OF THE LIVER

Fill in the blanks.

The liver plays an important role in the mechanical digestion of lipids because it secretes (1) ________________. It also produces two of the plasma proteins that play an essential role in blood clotting. These two proteins are (2) ______________ and (3) ______________. Additionally liver cells store several substances, notably vitamins A, B_{12}, D, and (4) ____________.

Finally, the liver is assisted by a unique structural feature of the blood vessels that supply it. This arrangement, known as the (5) ______________, allows toxins to be removed from the bloodstream before nutrients are distributed throughout the body.

NUTRIENT METABOLISM

Match the term with the definition.

(a) Carbohydrate (b) Fat (c) Protein
(d) Vitamins (e) Minerals

_____6. Used if cells have inadequate amounts of glucose to catabolize
_____7. Preferred energy food
_____8. Amino acids
_____9. Fat soluble
_____10. Required for nerve conduction
_____11. Glycolysis
_____12. Inorganic elements found naturally in the earth
_____13. Pyruvic acid

Circle the word or phrase that does not belong.

14. Glycolysis	Citric acid cycle	ATP	Bile
15. Adipose	Amino acids	Energy	Glycerol
16. A	D	M	K
17. Pyruvic acid	Proteins	Amino acids	Energy
18. Hydrocortisone	Insulin	Growth hormone	Epinephrine
19. Sodium	Calcium	Zinc	Folic acid
20. Thiamine	Niacin	Ascorbic acid	Riboflavin

▸ If you have had difficulty with this section, review pages 320-324. ◂

METABOLIC RATES
BODY TEMPERATURE

Circle the correct choice.

21. The rate at which food is catabolized under basal conditions is the:

a. TMR c. BMR
b. PBI d. ATP

22. The total amount of energy used by the body per day is the:

a. TMR c. BMR
b. PBI d. ATP

23. Over __________ of the energy released from food molecules during catabolism is converted to heat rather than being transferred to ATP.

a. 20%
b. 40%
c. 60%
d. 80%

24. Maintaining thermoregulation is a function of the:

a. Thalamus
b. Hypothalamus
c. Thyroid
d. Parathyroids

25. A transfer of heat energy to the skin, then the external environment is:

a. Radiation
b. Conduction
c. Convection
d. Evaporation

26. A flow of heat waves away from the blood is known as:

a. Radiation
b. Conduction
c. Convection
d. Evaporation

27. A transfer of heat energy to air that is continually flowing away from the skin is known as:

a. Radiation
b. Conduction
c. Convection
d. Evaporation

28. Heat that is absorbed by the process of water evaporation is called:

a. Radiation
b. Conduction
c. Convection
d. Evaporation

29. A/an ______________ is the amount of energy needed to raise the temperature of one gram of water one degree Celsius.

a. Calorie
b. Kilocalorie
c. ATP
d. BMR

▸ If you have had difficulty with this section, review pages 325-328. ◂

Unscramble the words.

30. LRIEV

31. TAOBALICMS

32. OMNIA

33. YPURCVI

Take the circled letters, unscramble them, and fill in the statement.

How the magician paid his bills.

34.

APPLYING WHAT YOU KNOW

35. Dr. Ellis was concerned about Deborrah. Her daily food intake provided fewer calories than her TMR. If this trend continues, what will be the result? If it continues over a long period of time, what eating disorder might Deborrah develop?

36. Lisa Kennedy was experiencing fatigue and a blood test revealed that she was slightly anemic. What mineral will her doctor most likely prescribe? What dietary sources might you suggest that she emphasize in her daily intake?

37. WORD FIND

Can you find the terms listed below in the box of letters? Words may be spelled top to bottom, bottom to top, right to left, left to right, or diagonally.

```
M I N E R A L S N K Y O H N K N I A L P
W B G K C F B M Z M C O N V E C T I O N
P J Z C V A H C I B P P J J R M T Z A I
B P U S X J R S I X O J O A E S L J C I
R T C D G R X B Y R A D I A T I O N A E
S A B I B O B U O H S A A V V F Y V T X
M Y A K Y Q L C W H C A E Y U F V L A K
Q E U P N L K U N G Y P Y N M S Q K B V
N T F Q Y F G R V R N D L H O R U G O X
O U E A F T X M T B L O R E C Y L G L H
I T H P T V B Q B M R Q C A F G V I I D
T X A Q G S U I Y R E V I L T D Y J S F
C L J O Y H F V T W I E Q G K E Z J M U
U Z C B C P O W F I A I N W E B S F Z D
D S N I E T O R P V Z R V G B Q P N C H
N J Y K J Z E Z Y A D I P O S E P Y J P
O K V X L V Q N N O I T A R O P A V E N
C Y O Z S Z Q M O F E L I B Z S U J Y G
A F Q H M S I S Y L O C Y L G P F H E K
O D I A A Q R Q V I T A M I N S U A U Q
```

LIST OF WORDS

LIVER
PROTEINS
GLYCOLYSIS
ADIPOSE
PBI
RADIATION
EVAPORATION
CARBOHYDRATES
VITAMINS
ATP
GLYCEROL
BMR
CONDUCTION
FATS
MINERALS
BILE
CATABOLISM
TMR
CONVECTION

DID YOU KNOW?

The amount of energy required to raise a 200-pound man 15 feet is about the amount of energy in one large calorie.

NUTRITION/METABOLISM

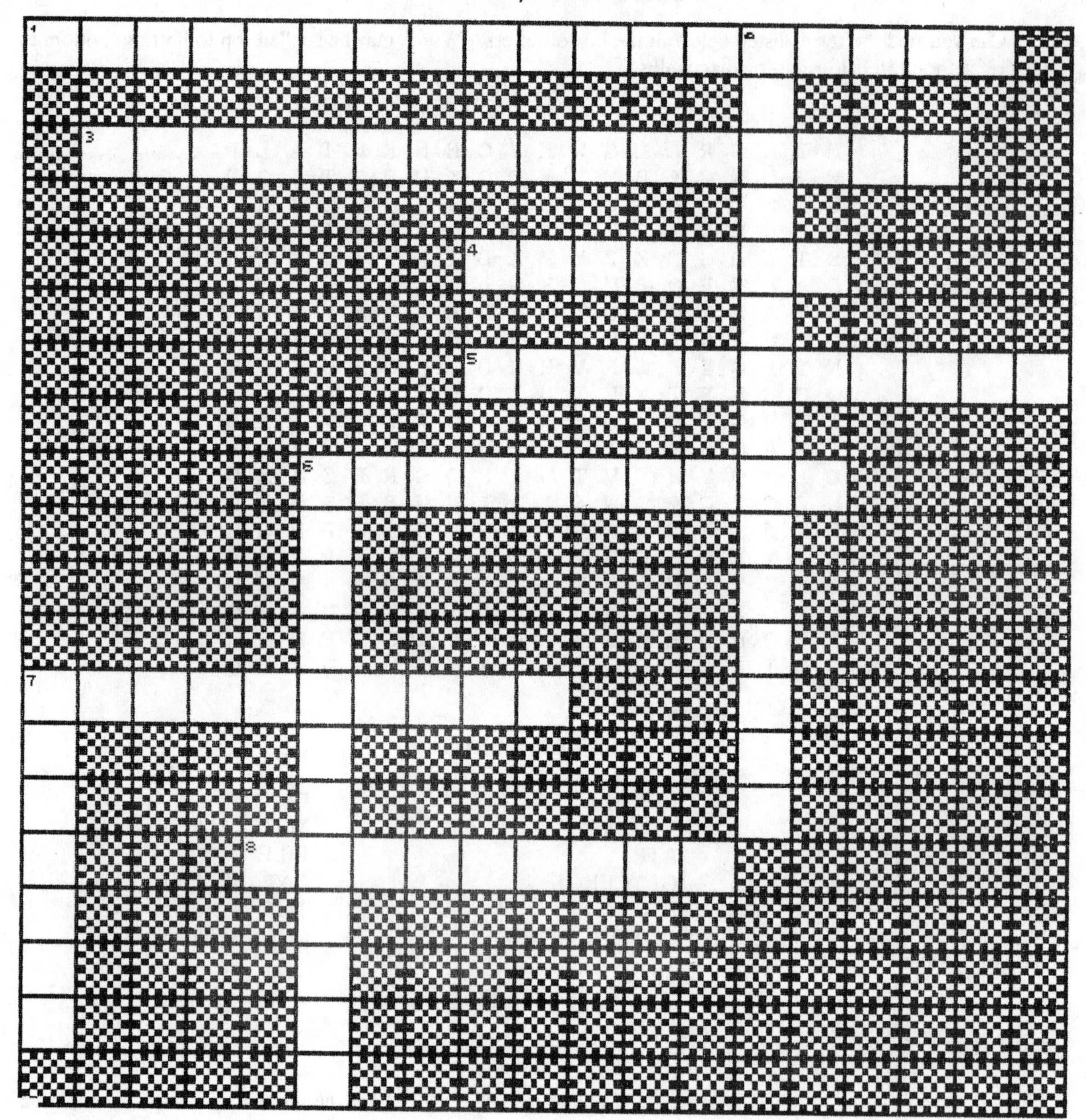

ACROSS

1. Rate of metabolism when a person is lying down, but awake (three words)
3. Maintaining homeostasis of body temperature
4. Organic molecule needed in small quantities for normal metabolism throughout the body
5. A unit of measure for heat, also known as a large calorie
6. Takes place in the cytoplasm of a cell and changes glucose to pyruvic acid
7. Breaks food molecules down releasing stored energy
8. Builds food molecules into complex substances

DOWN

2. An aerobic process which changes pyruvic acid to carbon dioxide
6. A series of reactions that join glucose molecules together to form glycogen
7. Amount of energy needed to raise the temperature of one gram of water one degree Celsius

CHAPTER 16

The Urinary System

Living produces wastes. Wherever people live or work or play, wastes accumulate. To keep these areas healthy, there must be a method of disposing of these wastes such as a sanitation department.

Wastes accumulate in your body also. The conversion of food and gases into substances and energy necessary for survival results in waste products. A large percentage of these wastes is removed by the urinary system.

Two vital organs, the kidneys, cleanse the blood of the many waste products that are continually produced as a result of the metabolism of food in the body cells. They eliminate these wastes in the form of urine.

Urine formation is the result of three processes: filtration, reabsorption, and secretion. These processes occur in successive portions of the microscopic units of the kidneys known as nephrons. The amount of urine produced by the nephrons is controlled primarily by two hormones, ADH and aldosterone.

After urine is produced it is drained from the renal pelvis by the ureters to flow into the bladder. The bladder then stores the urine until it is voided through the urethra.

If waste products are allowed to accumulate in the body, they soon become poisonous, a condition called uremia. A knowledge of the urinary system is necessary to understand how the body rids itself of waste and avoids toxicity.

TOPICS FOR REVIEW

Before progressing to Chapter 17 you should have an understanding of the structure and function of the organs of the urinary system. Your review should include knowledge of the nephron and its role in urine production. Your study should conclude with a review of the three main processes involved in urine production and the mechanisms that control urine volume.

KIDNEYS

Circle the correct choice.

1. The outermost portion of the kidney is known as the:

   a. Medulla
   b. Papilla
   c. Pelvis
   d. Pyramid
   e. Cortex

2. The saclike structure that surrounds the glomerulus is the:

   a. Renal pelvis
   b. Calyx
   c. Bowman's capsule
   d. Cortex
   e. None of the above is correct

3. The renal corpuscle is made up of the:

   a. Bowman's capsule and proximal convoluted tubule
   b. Glomerulus and proximal convoluted tubule
   c. Bowman's capsule and the distal convoluted tubule
   d. Glomerulus and the distal convoluted tubule
   e. Bowman's capsule and the glomerulus

4. Which of the following functions is *not* performed by the kidneys?

   a. Help maintain homeostasis
   b. Remove wastes from the blood
   c. Produce ADH
   d. Remove electrolytes from the blood

5. ______________________% of the glomerular filtrate is reabsorbed.

   a. 20
   b. 40
   c. 75
   d. 85
   e. 99

6. The glomerular filtration rate is ______________ ml per minute.

   a. 1.25
   b. 12.5
   c. 125.0
   d. 1250.0
   e. None of the above is correct

7. Glucose is mostly reabsorbed in the:

   a. Loop of Henle
   b. Proximal convoluted tubule
   c. Distal convoluted tubule
   d. Glomerulus
   e. None of the above is correct.

8. Reabsorption does *not* occur in the:

   a. Loop of Henle
   b. Proximal convoluted tubule
   c. Distal convoluted tubule
   d. Collecting tubules
   e. Calyx

9. The greater the amount of salt intake the:

   a. Less salt excreted in the urine
   b. More salt is reabsorbed
   c. The more salt excreted in the urine
   d. None of the above is correct.

10. Which one of the following substances is secreted by diffusion:

   a. Sodium ions
   b. Certain drugs
   c. Ammonia
   d. Hydrogen ions
   e. Potassium ions

11. Which of the following statements about ADH is *not* correct?

    a. It is stored by the pituitary gland
    b. It makes the collecting tubules less permeable to water
    c. It makes the distal convoluted tubules more permeable
    d. It is produced by the hypothalamus

12. Which of the following statements about aldosterone is *not* correct?

    a. It is secreted by the adrenal cortex
    b. It is a water-retaining hormone
    c. It is a salt-retaining hormone
    d. All of the above are correct

Choose the correct term and write its letter in the space next to the appropriate definition below.

a. Medulla
b. Cortex
c. Pyramids
d. Papilla
e. Pelvis
f. Calyx
g. Nephrons
h. Uremia
i. Proteinuria
j. Bowman's capsule
k. Glomerulus
l. Loop of Henle
m. CAPD
n. Glycosuria

_____13. Functioning unit of urinary system
_____14. Abnormally large amounts of plasma proteins in the urine
_____15. Uremic poisoning
_____16. Outer part of kidney
_____17. Together with Bowman's capsule forms renal corpuscle
_____18. Division of the renal pelvis
_____19. Cup-shaped top of a nephron
_____20. Innermost end of a pyramid
_____21. Extension of proximal tubule
_____22. Triangular-shaped divisions of the medulla of the kidney
_____23. Used in the treatment of renal failure
_____24. Inner portion of kidney

▸ If you have had difficulty with this section, review pages 332-341. ◂

URETERS
URINARY BLADDER
URETHRA

Indicate which organ is identified by the following descriptions by inserting the appropriate letter in the answer blank.

(a) Ureters (b) Bladder (c) Urethra

_____25. Rugae
_____26. Lower-most part of urinary tract
_____27. Lining membrane richly supplied with sensory nerve endings
_____28. Lies behind pubic symphysis

____29. Dual function in male
____30. 1 1/2 inches long in female
____31. Drains renal pelvis
____32. Surrounded by prostate in male
____33. Elastic fibers and involuntary muscle fibers
____34. 10 to 12 inches long
____35. Trigone

Fill in the blanks.

36. __________ __________ is the description of the pain caused by the passage of a kidney stone.

37. The urinary tract is lined with __________ __________.

38. Another name for kidney stones is __________ __________.

39. A technique that uses __________ to pulverize stones, thus avoiding surgery, is being used to treat kidney stones.

40. The passage of a tube through the urethra into the bladder for the removal of urine is known as __________.

41. A bladder infection may be referred to as __________.

42. In the male, the urethra serves a dual function: a passageway for urine and __________.

43. The external opening of the urethra is the __________ __________.

▸ If you have had difficulty with this section, review pages 342-345. ◂

MICTURITION

Fill in the blanks.

The terms (44) __________, (45) __________ and (46) __________ all refer to the passage of urine from the body or the emptying of the bladder. The sphincters guard the bladder. The (47) __________ __________ sphincter is located at the bladder (48) __________ and is involuntary. The external urethral sphincter circles the (49) __________ and is under (50) __________ control.

As the bladder fills, nervous impulses are transmitted to the spinal cord and an (51) __________ __________ is initiated. Urine then enters the (52) __________ to be eliminated.

Urinary (53) __________ is a condition in which no urine is voided. Urinary (54) __________ is when the kidneys do not produce any urine, but the bladder retains its ability to empty itself. Complete destruction or transection of the sacral cord produces an (55) __________ __________.

▸ If you have had difficulty with this section, review pages 344-345. ◂

APPLYING WHAT YOU KNOW

56. John suffered from low levels of ADH. What primary urinary symptom would he notice?

57. Bud was in a diving accident and his spinal cord was severed. He was paralyzed from the waist down and as a result was incontinent. His physician was concerned about the continuous residual urine buildup. What was the reason for concern?

58. **WORD FIND**

Can you find the terms listed below in the box of letters? Words may be spelled top to bottom, bottom to top, right to left, left to right, or diagonally.

```
B X Z X Z M E T H J Z G I R S C F U V X
F W N E P H R O N S I T I T S Y C K Y P
U Z B X Y Q E I A L L U D E M L D R E O
R J T M D P E R Q M C O R T E X J L N K
E S O W E B X D S E P H L X A Y U R D M
T U U L Q N S A X R E V A F B O O E I Z
E Z V H R X Z N K M G J H A G W J D K U
R I J L D H E Z O F I L T R A T I O N B
S H F J L K L D X T C P O N D Y D R B T
B P X I U E I Q Z W D Y X M O M I E M J
T J S Y H A A D H B P R U I E A I D S V
W F Z H L J O I N A N A C P B R S D B C
L D Q Y X A Q L P C Q M Q A W E U A C F
H Q S W C U C I R C G I P A L Y L L J X
R I W F Y B L E K E R D V S L C D B U I
S C F C H L O R N I A S Y C U S U Q L S
M R S T A Y Z K L M S V J I U A H L Q H
B K Q K Z N M T X R I H T K P I S V I D
J G D B E C N E N I T N O C N I P V E M
V A Z M I C T U R I T I O N L V I J I H
```

LIST OF WORDS

MEDULLA	CORTEX	PYRAMIDS
PAPILLA	PELVIS	CALYX
GLOMERULUS	CALCULI	INCONTINENCE
FILTRATION	NEPHRON	HEMODIALYSIS
ADH	URETERS	BLADDER
KIDNEY	MICTURITION	CYSTITIS

DID YOU KNOW?

If the tubules in a kidney were stretched out and untangled, there would be 70 miles of them.

URINARY SYSTEM

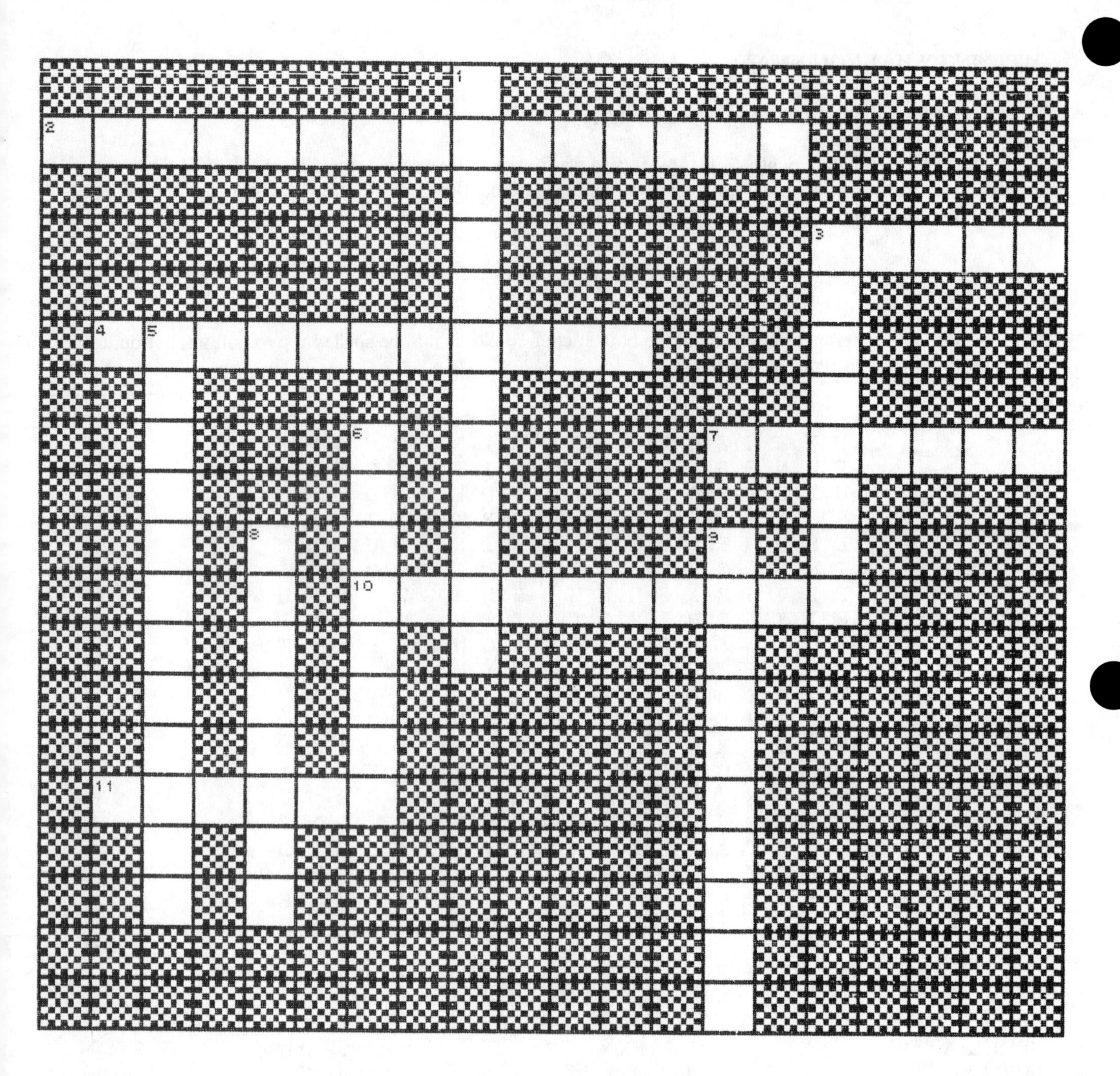

<u>ACROSS</u>

2. Passage of a tube into the bladder to withdraw urine
3. Division of the renal pelvis
4. Urination
7. Area on posterior bladder wall free of rugae
10. Network of blood capillaries tucked into Bowman's capsule
11. Absence of urine

<u>DOWN</u>

1. Ultrasound generator used to break up kidney stones
3. Bladder infection
5. Voiding involuntarily
6. Scanty urine
8. Large amount of urine
9. Glucose in the urine

URINARY SYSTEM

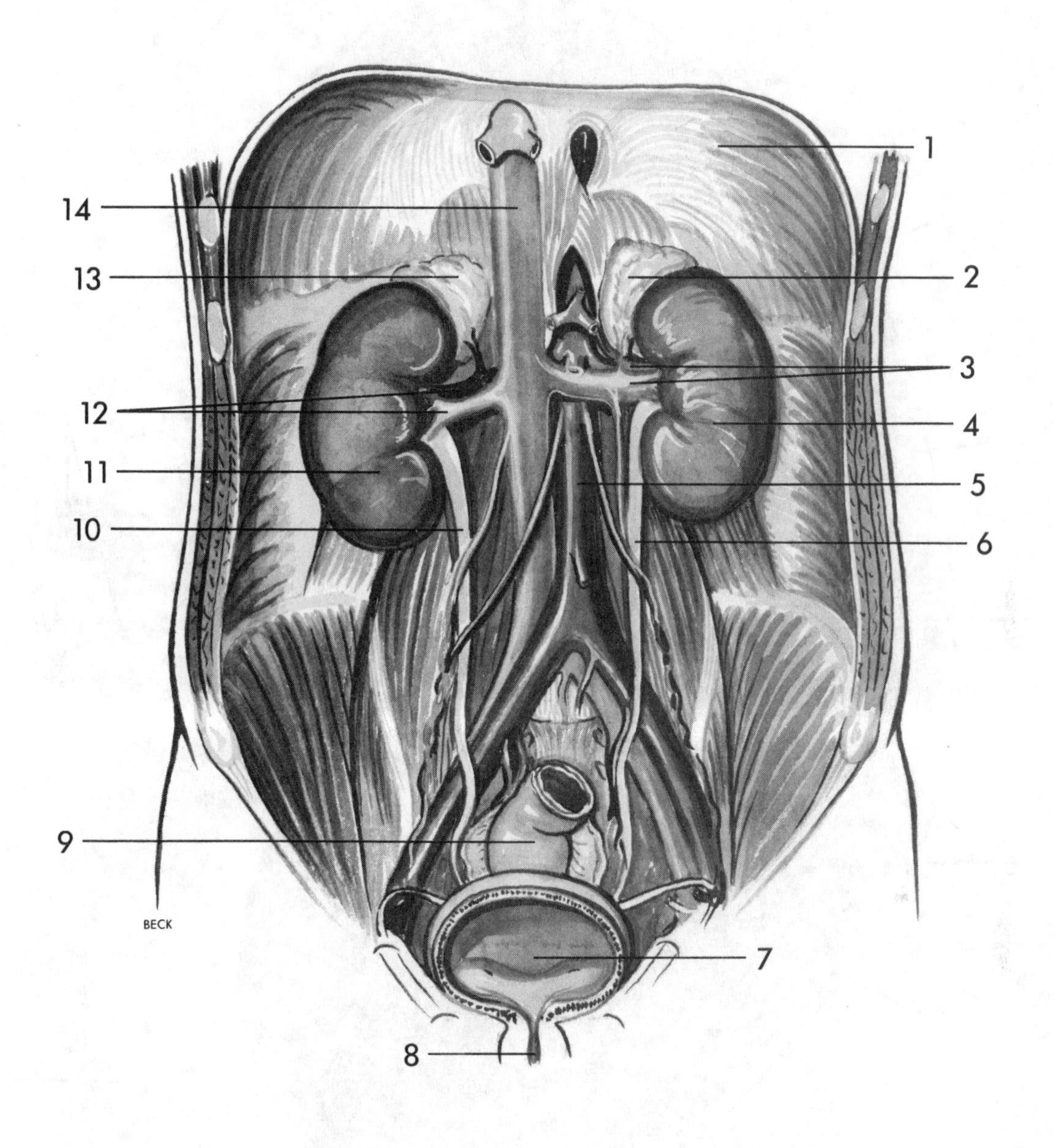

1.	______________	8.	______________
2.	______________	9.	______________
3.	______________	10.	______________
4.	______________	11.	______________
5.	______________	12.	______________
6.	______________	13.	______________
7.	______________	14.	______________

KIDNEY

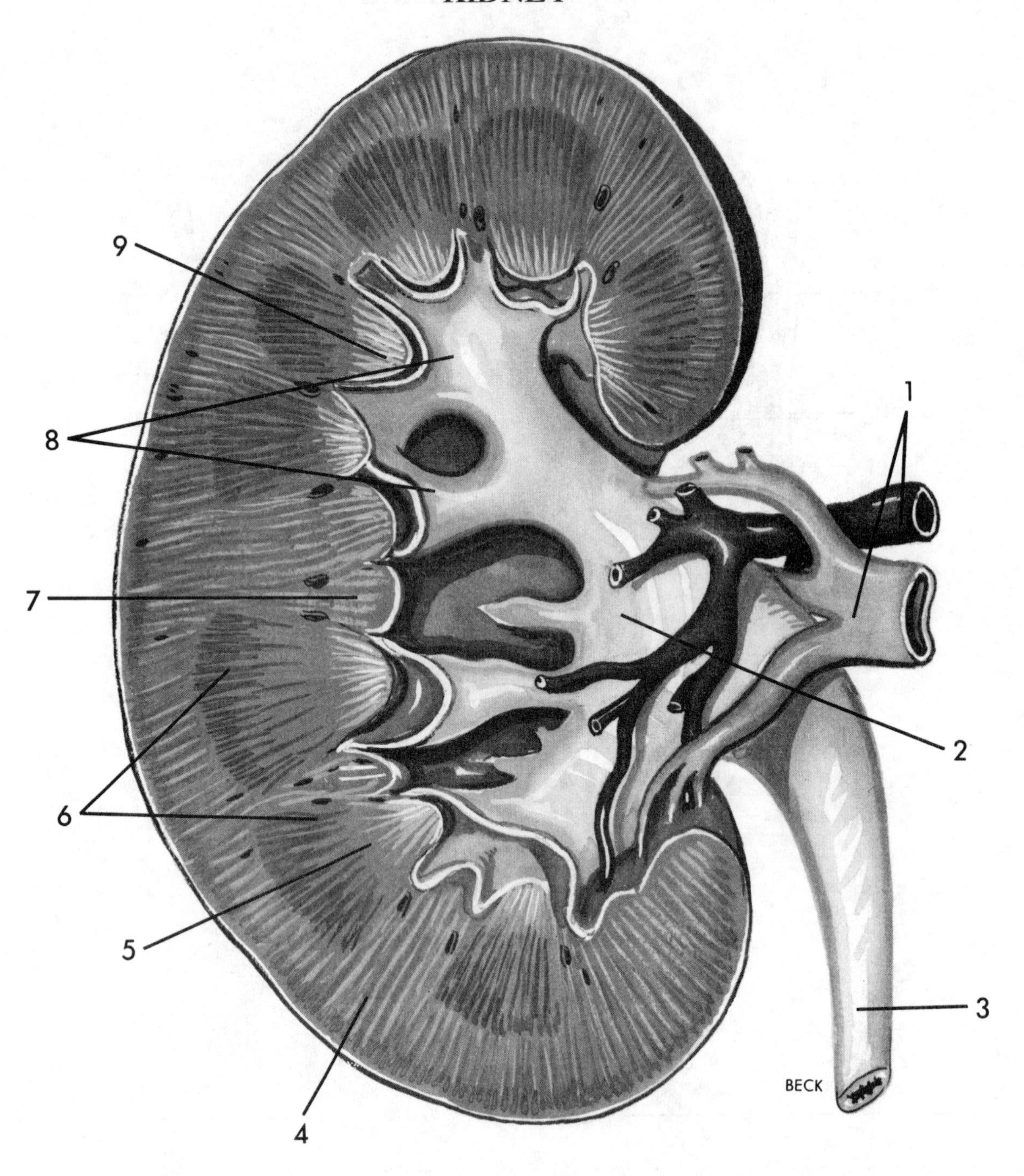

1. ______________________
2. ______________________
3. ______________________
4. ______________________
5. ______________________
6. ______________________
7. ______________________
8. ______________________
9. ______________________

NEPHRON

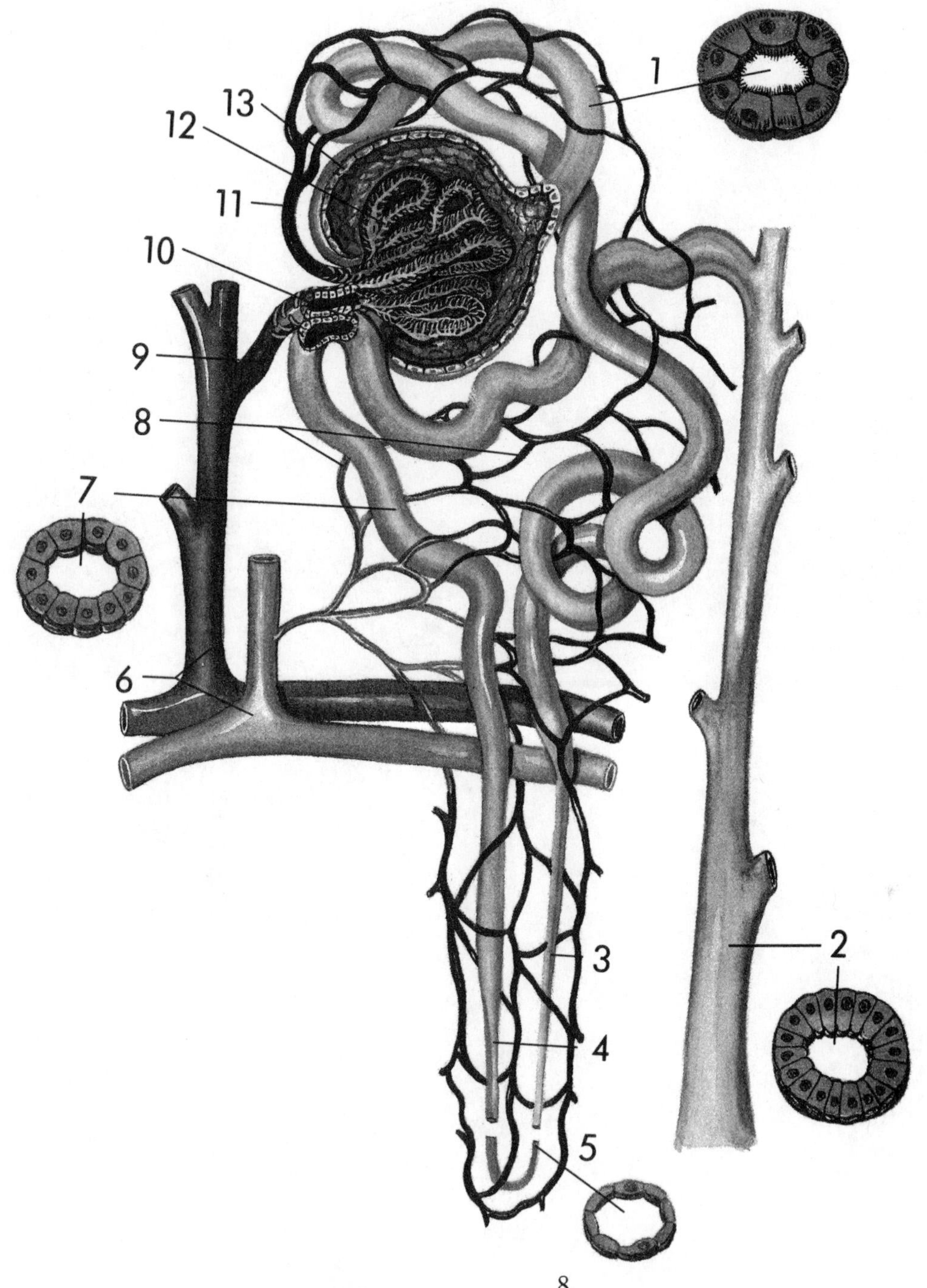

1. ______________________

2. ______________________

3. ______________________

4. ______________________

5. ______________________

6. ______________________

7. ______________________

8. ______________________

9. ______________________

10. ______________________

11. ______________________

12. ______________________

13. ______________________

CHAPTER 17

Fluid and Electrolyte Balance

Referring to the very first chapter in your text, you will recall that survival depends on the body's ability to maintain or restore homeostasis. Specifically, homeostasis means that the body fluids remain constant within very narrow limits. These fluids are classified as either intracellular fluid (ICF) or extracellular fluid (ECF). As their names imply, intracellular fluid lies within the cells and extracellular fluid is located outside the cells. A balance between these two fluids is maintained by certain body mechanisms. They are: (a) adjustment of fluid output to fluid intake under normal circumstances; (b) the concentration of electrolytes in the extracellular fluid; (c) the capillary blood pressure, and finally (d) the concentration of proteins in the blood.

Comprehension of how these mechanisms maintain and restore fluid balance is necessary for an understanding of the complexities of homeostasis and its relationship to the survival of the individual.

TOPICS FOR REVIEW

Before progressing to Chapter 18, you should review the types of body fluids and their subdivisions. Your study should include the mechanisms that maintain fluid balance and the nature and importance of electrolytes in body fluids. You should be able to give examples of common fluid imbalances, and have an understanding of the role of fluid and electrolyte balance in the maintenance of homeostasis.

BODY FLUIDS

Circle the correct answer.

1. The largest volume of water by far lies (inside or outside) cells.
2. Interstitial fluid is (intracellular or extracellular).
3. Plasma is (intracellular or extracellular).
4. Fat people have a (lower or higher) water content per pound of body weight than thin people.
5. Infants have (more or less) water in comparison to body weight than adults of either sex.
6. There is a rapid (increase or decline) in the proportion of body water to body weight during the first year of life.
7. The female body contains slightly (more or less) water per pound of weight.
8. In general, as age increases, the amount of water per pound of body weight (increases or decreases).
9. Excluding adipose tissue, approximately (55% or 85%) of body weight is water.
10. The term (fluid balance or fluid compartments) means the volumes of ICF, IF, plasma, and the total volume of water in the body all remain relatively constant.

▸ If you have had difficulty with this section, review pages 350-351. ◂

MECHANISMS THAT MAINTAIN FLUID BALANCE

Circle the correct choice.

11. Which one of the following is *not* an anion?

 a. Chloride
 b. Bicarbonate
 c. Sodium
 d. Many proteins

12. Which one of the following is *not* a cation?

 a. Sodium
 b. Potassium
 c. Calcium
 d. Magnesium
 e. All of the above are cations

13. The most abundant cation in the blood plasma is:

 a. Sodium
 b. Chloride
 c. Protein
 d. Calcium
 e. Magnesium

14. If the blood sodium concentration increases, then blood volume will:

 a. Increase
 b. Decrease
 c. Remain the same
 d. None of the above is correct

15. The smallest amount of water comes from:

 a. Water in foods that are eaten
 b. Ingested liquids
 c. Water formed from catabolism
 d. None of the above is correct

16. The greatest amount of water lost from the body is from the:

 a. Lungs
 b. Skin by diffusion
 c. Skin by sweat
 d. Feces
 e. Kidneys

17. Which one of the following is *not* a major factor that influences extracellular and intracellular fluid volumes?

 a. The concentration of electrolytes in the extracellular fluid
 b. The capillary blood pressure
 c. The concentration of proteins in blood
 d. All of the above are important factors

18. The type of fluid output that changes the most is:

 a. Water loss in the feces
 b. Water loss across the skin
 c. Water loss via the lungs
 d. Water loss in the urine
 e. None of the above is correct

19. The chief regulators of sodium within the body are the:

a. Lungs
b. Sweat glands
c. Kidneys
d. Large intestine
e. None of the above is correct

20. Which of the following is *not* correct?

a. Fluid output must equal fluid intake.
b. ADH controls salt reabsorption in the kidney.
c. Water follows sodium.
d. Renal tubule regulation of salt and water is the most important factor in determining urine volume.
e. All of the above are correct.

21. Diuretics work on all but which one of the following?

a. Proximal tubule
b. Loop of Henle
c. Distal tubule
d. Collecting ducts
e. Diuretics work on all of the above

22. Of all the sodium-containing secretions, the one with the largest volume is:

a. Saliva
b. Gastric secretions
c. Bile
d. Pancreatic juice
e. Intestinal secretions

23. The higher the capillary blood pressure, the ___________ the amount of interstitial fluid.

a. Smaller
b. Larger
c. There is no relationship between capillary blood pressure and volume of interstitial fluid

24. An increase in capillary blood pressure will lead to ___________ in blood volume.

a. An increase
b. A decrease
c. No change
d. None of the above is correct

25. Which one of the fluid compartments varies the most in volume?

a. Intracellular
b. Interstitial
c. Extracellular
d. Plasma

26. Which one of the following will *not* cause edema?

a. Retention of electrolytes in the extracellular fluid
b. Increase in capillary blood pressure
c. Burns
d. Decrease in plasma proteins
e. All of the above will cause edema

If the following statements are true, insert T in the answer blanks. If any of the statements are false, circle the incorrect word(s) and write the correct word in the answer blank.

_________ 27. The three sources of fluid intake are: the liquids we drink, the foods we eat, and water formed by the anabolism of foods.
_________ 28. The body maintains fluid balance mainly by changing the volume of urine excreted to match changes in the volume of fluid intake.
_________ 29. Some output of fluid will occur as long as life continues.
_________ 30. Glucose is an example of an electrolyte.
_________ 31. Where sodium goes, water soon follows.
_________ 32. Excess aldosterone leads to hypovolemia.
_________ 33. Diuretics have their effect on glomerular function.
_________ 34. Typical daily intake and output totals should be approximately 1200 ml.
_________ 35. Bile is a sodium-containing internal secretion.
_________ 36. The average daily diet contains about 500 mEg of sodium.

▸ If you have had difficulty with this section, review pages 352-359. ◂

FLUID IMBALANCES

Fill in the blanks.

(37)_______________ is the fluid imbalance seen most often. In this condition, interstitial fluid volume (38) _______________ first, but eventually, if treatment has not been given, intracellular fluid and plasma volumes (39) _________________. (40) _________________ can also occur, but is much less common. Giving (41) _______________ too rapidly or in too large amounts can put too heavy a burden on the (42) _______________.

▸ If you have had difficulty with this section, review page 359. ◂

APPLYING WHAT YOU KNOW

43. Mrs. Titus was asked to keep an accurate record of her fluid intake and output. She was concerned because the two did not balance. What is a possible explanation for this?

44. Nurse Briker was caring for a patient who was receiving diuretics. What special nursing implications should be followed for patients on this therapy?

FLUID/ELECTROLYTES

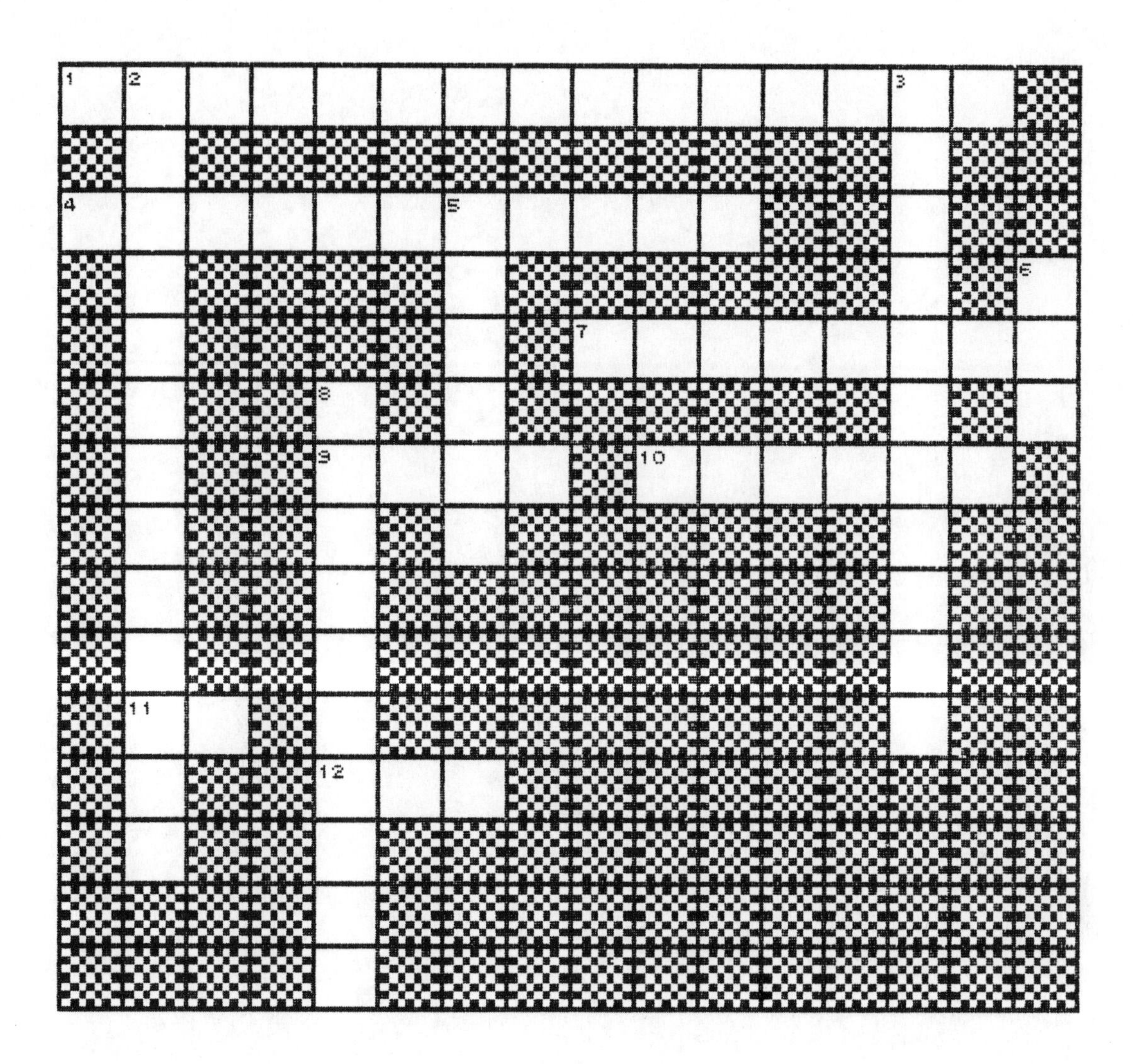

ACROSS

1. Organic substances that don't dissociate in solution
4. Total volume of body fluids less than normal
7. "Causing urine"
9. Dissociated particles of an electrolyte that carry an electrical charge
10. Positively charged ion
11. A subdivision of extracellular fluid (abbrev.)
12. Fluid inside cells (abbrev.)

DOWN

2. Total volume of body fluids greater than normal
3. Compound that dissociates in solution into ions
5. Negatively charged ions
6. Fluid outside cells (abbrev.)
8. To break up

CHAPTER 18

Acid-Base Balance

It has been established in previous chapters that an equilibrium between intracellular and extracellular fluid volume must exist for homeostasis. Equally important to homeostasis is the chemical acid-base balance of the body fluids. The degree of acidity or alkalinity of a body fluid is expressed in pH value. The neutral point, where a fluid would be neither acid nor alkaline, is pH 7. Increasing acidity is expressed as less than 7, and increasing alkalinity as greater than 7. Examples of body fluids that are acidic are gastric juice (1.6) and urine (6.0). Blood, on the other hand, is considered alkaline with a pH of 7.45.

Buffers are substances that prevent a sharp change in the pH of a fluid when an acid or base is added to it. They are one of several mechanisms that are constantly monitoring the pH of fluids in the body. If, for any reason, these mechanisms do not function properly, a pH imbalance occurs. These two kinds of imbalances are known as alkalosis and acidosis.

Maintaining the acid-base balance of body fluids is a matter of vital importance. If this balance varies even slightly, necessary chemical and cellular reactions cannot occur. Your review of this chapter is necessary to understand the delicate fluid balance necessary to survival.

TOPICS FOR REVIEW

Before progressing to Chapter 19 you should have an understanding of the pH of body fluids and the mechanisms that control the pH of these fluids in the body. Your study should conclude with a review of the metabolic and respiratory types of pH imbalances.

pH OF BODY

Write the letter of the correct term on the blank next to the appropriate statement.

(a) Acid (b) Base

_____1. Lower concentration of hydrogen ions than hydroxide ions
_____2. Higher concentration of hydrogen ions than hydroxide ions
_____3. Gastric juice
_____4. Saliva
_____5. Arterial blood
_____6. Venous blood
_____7. Baking soda
_____8. Milk
_____9. Ammonia
_____10. Egg white

▸ If you have had difficulty with this section, review pages 364-365. ◂

MECHANISMS THAT CONTROL pH OF BODY FLUIDS

Circle the correct choice.

11. When carbon dioxide enters the blood it reacts with the enzyme carbonic anhydrase to form:

    a. Sodium bicarbonate
    b. Water and carbon dioxide
    c. Ammonium chloride
    d. Bicarbonate ion
    e. Carbonic acid

12. The lungs remove ______________ liters of carbonic acid each day.

    a. 10.0
    b. 15.0
    c. 20.0
    d. 25.0
    e. 30.0

13. When a buffer reacts with a strong acid it changes the strong acid to a:

    a. Weak acid
    b. Strong base
    c. Weak base
    d. Water
    e. None of the above is correct

14. Which one of the following is *not* a change in the blood that results from the buffering of fixed acids in tissue capillaries?

    a. The amount of carbonic acid increases slightly.
    b. The amount of bicarbonate in blood decreases.
    c. The hydrogen ion concentration of blood increases slightly.
    d. The blood pH decreases slightly.
    e. All of the above are changes that result from the buffering of nonvolatile acids in tissue capillaries.

15. The most abundant acid in the body is:

    a. HCL
    b. Lactic acid
    c. Carbonic acid
    d. Acetic acid
    e. Sulfuric acid

16. The normal ratio of sodium bicarbonate to carbonic acid in arterial blood is:

    a. 5:1
    b. 10:1
    c. 15:1
    d. 20:1
    e. None of the above is correct

17. Which of the following would *not* be a consequence of holding your breath?

    a. The amount of carbonic acid in the blood would increase.
    b. The blood pH would decrease.
    c. The body would develop an alkalosis.
    d. No carbon dioxide could leave the body.

18. Which of the following is *not* true of the kidneys?

a. They can eliminate larger amounts of acid than the lungs.
b. More bases than acids are usually excreted by the kidneys.
c. If the kidneys fail, homeostasis of acid-base balance fails.
d. They are the most effective regulators of blood pH.

19. The pH of the urine may be as low as:

a. 1.6
b. 2.5
c 3.2
d. 4.8
e. 7.4

20. In the distal tubule cells the product of the reaction aided by carbonic anhydrase is:

a. Water
b. Carbon dioxide
c. Water and carbon dioxide
d. Hydrogen ions
e. Carbonic acid

21. In the distal tubule, _______________ leaves the tubule cells and enters the blood capillaries.

a. Carbon dioxide
b. Water
c. HCO_3
d. NaH_2PO_4
e. $NaHCO_3$

Mark T in the answer blank if the statement is true. If the statement is false, circle the incorrect word(s) and correct the statement on the answer blank.

_________ 22. The body has three mechanisms for regulating the pH of its fluids. They are the heart mechanism, the respiratory mechanism, and the urinary mechanism.
_________ 23. Buffers consist of two kinds of substances and are therefore often called duobuffers.
_________ 24. Oranges and grapefruit are not acid forming when metabolized.
_________ 25. Some athletes have adopted a technique called bicarbonate loading, ingesting large amounts of sodium bicarbonate ($NaHCO_3$) to counteract the effects of lactic acid buildup.
_________ 26. Anything that causes an excessive increase in respiration will in time produce acidosis.
_________ 27. The lungs are the body's most effective regulator of blood pH.
_________ 28. More acids than bases are usually excreted by the kidneys because more acids than bases usually enter the blood.
_________ 29. Blood levels of sodium bicarbonate can be regulated by the lungs.
_________ 30. Blood levels of carbonic acid can be regulated by the kidneys.

▸ If you have had difficulty with this section, review pages 364-372. ◂

METABOLIC AND RESPIRATORY DISTURBANCES

Write the letter of the correct term on the blank next to the appropriate definition.

a. Metabolic acidosis
b. Metabolic alkalosis
c. Respiratory acidosis
d. Respiratory alkalosis
e. Vomiting
f. Normal saline
g. Uncompensated metabolic acidosis
h. Hyperventilation
i. Hypersalivation
j. Ipecac

_____31. Emesis
_____32. Result of untreated diabetes
_____33. Chloride containing solution
_____34. Bicarbonate deficit
_____35. Present during emesis
_____36. Bicarbonate excess
_____37. Rapid breathing
_____38. Carbonic acid excess
_____39. Carbonic acid deficit
_____40. Emetic

▸ If you have had difficulty with this section, review pages 372-373. ◂

APPLYING WHAT YOU KNOW

41. Cara was pregnant and was experiencing repeated vomiting episodes for several days. Her doctor became concerned, admitted her to the hospital, and began intravenous administrations of normal saline. How will this help Cara?

42. Holly had a minor bladder infection. She had heard that this is often the result of the urine being less acidic than necessary, and that she should drink cranberry juice to correct the acid problem. She had no cranberry juice, so she decided to substitute orange juice. What was wrong with this substitution?

43. Mr. Cameron has frequent bouts of hyperacidity of the stomach. Which will assist in neutralizing the acid more promptly; milk or milk of magnesia?

44. **WORD FIND**

Can you find the terms from this chapter listed below in the box of letters? Words may be spelled top to bottom, bottom to top, right to left, left to right, or diagonally.

```
Q G J F M B F N A K N H A C J L J F N W
W K D I U R E T I C Y Q Y I W Y I C O V
D B X Z X E C N A L A B D I U L F Z I M
E E T S H J A L D O S T E R O N E Z T G
H E I R N S C F U V X F W K P Z B X A Y
Y L Q E I O L D R O J T M D E R Q H R M
D E J L K S I O W S B X D S E P O L D X
R C A Y U R M T O U U Q K N S M A X Y R
A T V A F B O D A O E Z Z I E H R X H Z
T R N K J H I A G C W I J O D D U J R L
I O D H E U A Z B O H N S F J N L K E L
O L X T M N C N U D Y T D B T B E P V I
N Y U E I Q Z T W D A A X O M I D Y O M
J T T O J S P H B S U K T W I A E I S S
V E N W F U Z H I J O E S I A N M N P B
S S B C T L D S Q X Q L R C Q T A Q W E
C F J Q W C U R H C G P I A Y L E J X R
W F Y B E K E D R V S L H D I C F R C H
O R N I A Y A C U S Q L T M R S T Y Z K
L N O N E L E C T R O L Y T E S M S V J
```

LIST OF WORDS

FLUID BALANCE
OUTPUT
HOMEOSTASIS
SODIUM
ADH
CATIONS

ALDOSTERONE
KIDNEYS
ELECTROLYTES
DIURETIC
DEHYDRATION
ANIONS

INTAKE
WATER
THIRST
EDEMA
OVERHYDRATION
NONELECTROLYTES

ACID/BASE BALANCE

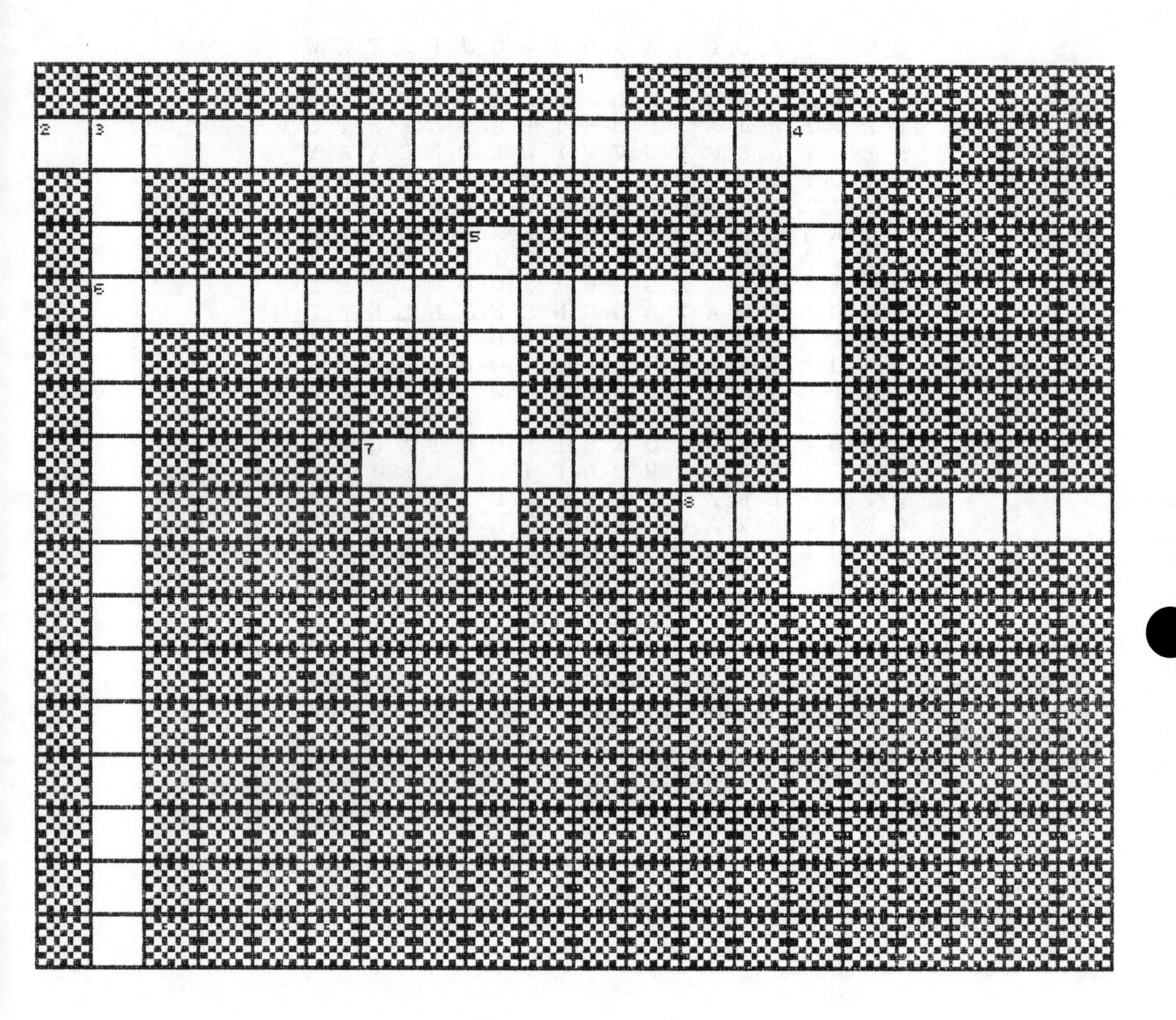

ACROSS

2. An enzyme found in red blood cells (two words)
6. Substance with a pH lower than 7.0 (two words)
7. Vomitus
8. Acid-base imbalance

DOWN

1. Way to express the acidity or alkalinity of a substance
3. Substance with a pH higher than 7.0 (two words)
4. Serious complication of vomiting
5. Prevents a sharp change in the pH of fluids

CHAPTER 19

The Reproductive Systems

The reproductive system consists of those organs that participate in perpetuating the species. It is a unique body system in that its organs differ between the two sexes, and yet the goal of creating a new being is the same. Of interest also is the fact that this system is the only one not necessary to the survival of the individual, and yet survival of the species depends on the proper functioning of the reproductive organs. The male reproductive system is divided into the external genitals: the testes, the duct system, and accessory glands. The testes, or gonads, are considered essential organs because they produce the sex cells, sperm, which join with the female sex cells, ova, to form a new human being. They also secrete the male sex hormone, testosterone, which is responsible for the physical transformation of a boy to a man.

Sperm are formed in the testes by the seminiferous tubules. From there they enter a long narrow duct, the epididymis. They continue onward through the vas deferens into the ejaculatory duct, down the urethra, and out of the body. Throughout this journey, various glands secrete substances that add motility to the sperm and create a chemical environment conducive to reproduction.

The female reproductive system is truly extraordinary and diverse. It produces ova, receives the penis and sperm during intercourse, is the site of conception, houses and nourishes the embryo during prenatal development, and nourishes the infant after birth.

Because of its diversity, the physiology of the female is generally considered to be more complex than that of the male. Much of the activity of this system revolves around the menstrual cycle and the monthly preparation that the female undergoes for a possible pregnancy.

The organs of this system are divided into essential organs and accessory organs of reproduction. The essential organs of the female are the ovaries. Just as with the male, the essential organs of the female are referred to as the gonads. The gonads of both sexes produce the sex cells. In the male, the gonads produce the sperm and in the female they produce the ova. The gonads are also responsible for producing the hormones in each sex necessary for the appearance of the secondary sex characteristics.

The menstrual cycle of the female typically covers a period of 28 days. Each cycle consists of three phases: the menstrual period, the postmenstrual phase, and the premenstrual phase. Changes in the blood levels of the hormones that are responsible for the menstrual cycle also cause physical and emotional changes in the female. A knowledge of these phenomena and this system, in both the male and the female, are necessary to complete your understanding of the reproductive system.

TOPICS FOR REVIEW

Before progressing to Chapter 20 you should familiarize yourself with the structure and function of the organs of the male and female reproductive systems. Your review should include emphasis on the gross and microscopic structure of the testes and the production of sperm and testosterone. Your study should continue by tracing the pathway of a sperm cell from formation to expulsion from the body.

You should then familiarize yourself with the structure and function of the organs of the female reproductive system. Your review should include emphasis on the development of a mature ova from ovarian follicles, and should additionally concentrate on the phases and occurrences in a typical 28-day menstrual cycle.

STRUCTURAL PLAN

Match the term on the left with the proper selection on the right.

Group A

_____1. Testes
_____2. Spermatozoa
_____3. Ova
_____4. Penis
_____5. Zygote

a. Fertilized ovum
b. Accessory organ
c. Male sex cell
d. Gonads
e. Gamete

Group B

_____6. Testes
_____7. Bulbourethral
_____8. Asexual
_____9. Erectile tissue
_____10. Prostate

a. Cowper's gland
b. Corpus cavernosum
c. Essential organ
d. Single parent
e. Accessory organ

▸ If you have had difficulty with this section, review pages 378-379. ◂

TESTES

Circle the correct choice.

11. The testes are surrounded by a tough membrane called:

a. Ductus deferens
b. Tunica albuginea
c. Septum
d. Seminiferous membrane

12. The ____________ lie near the septa that separate the lobules.

a. Ductus deferens
b. Sperm
c. Interstitial cells
d. Nerves

13. Sperm are found in the walls of the ____________.

a. Seminiferous tubule
b. Interstitial cells
c. Septum
d. Blood vessels

14. An undescended testicle is:

a. Orchidalgia
b. Orchidorrhaphy
c. Orchichorea
d. Cryptorchidism

15. The structure that produces testosterone is (are) the:

a. Seminiferous tubules
b. Prostate gland
c. Bulbourethral gland
d. Pituitary gland
e. Interstitial cells

16. The part of the sperm that contains genetic information that will be inherited is the:

a. Tail
b. Neck
c. Middle piece
d. Head
e. Acrosome

17. Which one of the following is *not* a function of testosterone?

a. It causes a deepening of the voice.
b. It promotes the development of the male accessory glands.
c. It has a stimulatory effect on protein catabolism.
d. It causes greater muscular development and strength.

18. Sperm production is called:

a. Spermatogonia
b. Spermatids
c. Spermatogenesis
d. Spermatocyte

19. The section of the sperm that contains enzymes that enable it to break down the covering of the ovum and permit entry should contact occur is the:

a. Acrosome
b. Midpiece
c. Tail
d. Stem

20. Descent of the testes usually occurs about ____________.

a. Two months after birth
b. Two months before birth
c. Two months after conception
d. Two years after birth
e. None of the above is correct

Fill in the blanks.

The (21) ____________ are the gonads of the male. From puberty on, the seminiferous tubules are continuously forming (22) ____________. Any of these cells may join with the female sex cell, the (23) ____________ to become a new human being.

Another function of the testes is to secrete the male hormone (24) ____________, that transforms a boy to a man. This hormone is secreted by the (25) ______________ of the testes. A good way to remember testosterone's functions is to think of it as "the (26) ______________ hormone" and "the (27) ______________ hormone."

▸ If you have had difficulty with this section, review pages 379-384. ◂

DUCTS
ACCESSORY MALE REPRODUCTIVE GLANDS

Choose the correct term and write its letter in the space next to the appropriate definition below.

a. Epididymis
b. Vas deferens
c. Ejaculatory duct
d. Prepuce
e. Seminal vesicles
f. Prostate gland
g. Cowper's gland
h. Prostatectomy
i. Semen
j. Scrotum

_____28. Continuation of ducts that start in epididymis
_____29. Procedure performed for benign prostatic hypertrophy
_____30. Also known as bulbourethral
_____31. Narrow tube that lies along the top and behind the testes
_____32. Doughnut-shaped gland beneath bladder
_____33. Continuation of vas deferens
_____34. Mixture of sperm and secretions of accessory sex glands
_____35. Contributes 60% of the seminal fluid volume
_____36. Removed during circumcision
_____37. External genitalia

▸ If you have had difficulty with this section, review pages 379-386. ◂

STRUCTURAL PLAN

Match the term on the left with the proper selection on the right.

_____38. Ovaries
_____39. Vagina
_____40. Bartholin
_____41. Vulva
_____42. Ova

a. Genitals
b. Accessory sex gland
c. Accessory duct
d. Gonads
e. Sex cell

Write the letter of the correct description in the blank next to the appropriate structure.

(a) External structure
(b) Internal structure

_____43. Mons pubis
_____44. Vagina
_____45. Labia majora
_____46. Uterine tubes
_____47. Vestibule
_____48. Clitoris
_____49. Labia minora
_____50. Ovaries

▸ If you have had difficulty with this section, review pages 389 and 392. ◂

OVARIES

Fill in the blanks.

The ovaries are the (51) _____________ of the female. They have two main functions. The first is the production of the female sex cell. This process is called (52) _____________. The specialized type of cell division that occurs during sexual cell reproduction is known as (53) _____________. The ovum is the body's largest cell and has (54) _____________ the number of chromosomes found in other body cells. At the time of (55) _____________, the sex cells from both parents fuse and (56) _____________ chromosomes are united.

The second major function of the ovaries is to secrete the sex hormones (57) _____________ and (58) _____________. Estrogen is the sex hormone that causes the development and maintenance of the female (59) _____________________. Progesterone acts with estrogen to help initiate the (60) _________________ in girls entering (61) ______________.

▸ If you have had difficulty with this section, review pages 386-388. ◂

FEMALE REPRODUCTIVE DUCTS

Write the letter of the correct structure in the blank next to the appropriate definition.

(a) Uterine tubes (b) Uterus (c) Vagina

_____62. Ectopic pregnancy
_____63. Lining known as endometrium
_____64. Terminal end of birth canal
_____65. Site of menstruation
_____66. Approximately 4 inches in length
_____67. Consists of body, fundus, and cervix
_____68. Site of fertilization
_____69. Also known as oviduct
_____70. Entrance way for sperm
_____71. Total hysterectomy

▸ If you have had difficulty with this section, review pages 388-390. ◂

ACCESSORY FEMALE REPRODUCTIVE GLANDS
EXTERNAL GENITALS OF THE FEMALE

Match the term on the left with the proper selection on the right.

Group A

_____72. Bartholin's gland
_____73. Breasts
_____74. Alveoli
_____75. Lactiferous ducts
_____76. Areola

a. Colored area around nipple
b. Grapelike clusters of milk-secreting cells
c. Drain alveoli
d. Secretes lubricating fluid
e. Primarily fat tissue

Group B

_____77. Mons pubis
_____78. Labia majora
_____79. Clitoris
_____80. Vestibule
_____81. Episiotomy

a. "Large lips"
b. Area between labia minora
c. Surgical procedure
d. Composed of erectile tissue
e. Pad of fat over the symphysis pubis

▸ If you have had difficulty with this section, review pages 390-392. ◂

MENSTRUAL CYCLE

If the following statement is true, insert T in the answer blank. If the statement is false, circle the incorrect word(s) and insert the correct word(s) in the answer blank.

_________82. Climacteric is the scientific name for the beginning of the menses.
_________83. As a general rule, several ovum mature each month during the 30-40 years that a woman has menstrual periods.
_________84. Ovulation occurs 28 days before the next menstrual period begins.
_________85. The first day of ovulation is considered the first day of the cycle.
_________86. A woman's fertile period lasts only a few days out of each month.
_________87. The control of the menstrual cycle lies in the posterior pituitary gland.

Write the letter of the correct hormone in the blank next to the appropriate description.

(a) FSH (b) LH

_____88. Ovulating hormone
_____89. Secreted during first days of menstrual cycle
_____90. Secreted after estrogen level of blood increases
_____91. Causes final maturation of follicle and ovum
_____92. Birth control pills suppress this one

▸ If you have had difficulty with this section, review pages 392-395. ◂

APPLYING WHAT YOU KNOW

93. Mr. Belinki is going into the hospital for the surgical removal of his testes. As a result of this surgery, will Mr. Belinki be sterile or impotent?

94. When baby Gaylor was born, the pediatrician discovered that his left testicle had not descended into the scrotum. If this situation is not corrected soon might baby Gaylor be sterile or impotent?

95. Marcia contracted gonorrhea. By the time she made an appointment to see her doctor, it had spread to her abdominal organs. How is this possible when gonorrhea is a disease of the reproductive system?

96. Mrs. Harlan was having a bilateral oophorectomy. Is this a sterilization procedure? Will she experience menopause?

97. Ms. Comstock had a total hysterectomy. Will she experience menopause?

98. **WORD FIND**

Can you find the terms from this chapter listed below in the box of letters? Words may be spelled top to bottom, bottom to top, right to left, left to right, or diagonally.

```
L K B W M H W K K L R Q Q P C W V U X X
G K P X U C U U J H Y Q X Z T B D L M M
A B I Y T I O V I D U C T S C K R R C U
C H A Y O M L P G K C F C O W P E R S M
R L V W R F Q I M X B M H Z T G Z O R S
O G X T C L J T Q A M V E A H L L G T I
S S S S S H P I C M O R H I B M F O X D
O N S E M I N I F E R O U S O S P D P I
M E I Q E N D O M E T R I U M S Z F S H
E R J K I F K U L J P P J H M R I Z I C
D E E G J Q O Q S P E R M A T I D S M R
E F W S L E S T R O G E N K E S T Y Y O
S E J E L S Q Z Q J J R I V Y P Y H D T
Q D A I C G X E L Y C M W R J E F V I P
W S P R O S T A T E C T O M Y R S A D Y
R A Q A C I P B F X F M N X I M C G I R
I V U V C E E E P V F E L Z T U A I P C
P M M O V R N Q L Y W U B U T B B N E E
E Q H A B X I N D G D S P I C D O A G O
L W Z X E Z S P R E G N A N C Y T O J S
```

LIST OF WORDS

SPERM	SEMINIFEROUS	SCROTUM
CRYPTORCHIDISM	MEIOSIS	SPERMATIDS
ACROSOME	EPIDIDYMIS	VASDEFERENS
COWPERS	PENIS	PROSTATECTOMY
OVARIES	ESTROGEN	OVIDUCTS
PREGNANCY	ENDOMETRIUM	VAGINA

DID YOU KNOW?

The ovaries contain approximately 500,000 ova cells, but only 400+ will mature and be capable of being fertilized.

The testes produce approximately 50 million sperm per day. Every 2 months they produce enough cells to populate the entire earth.

REPRODUCTIVE SYSTEM

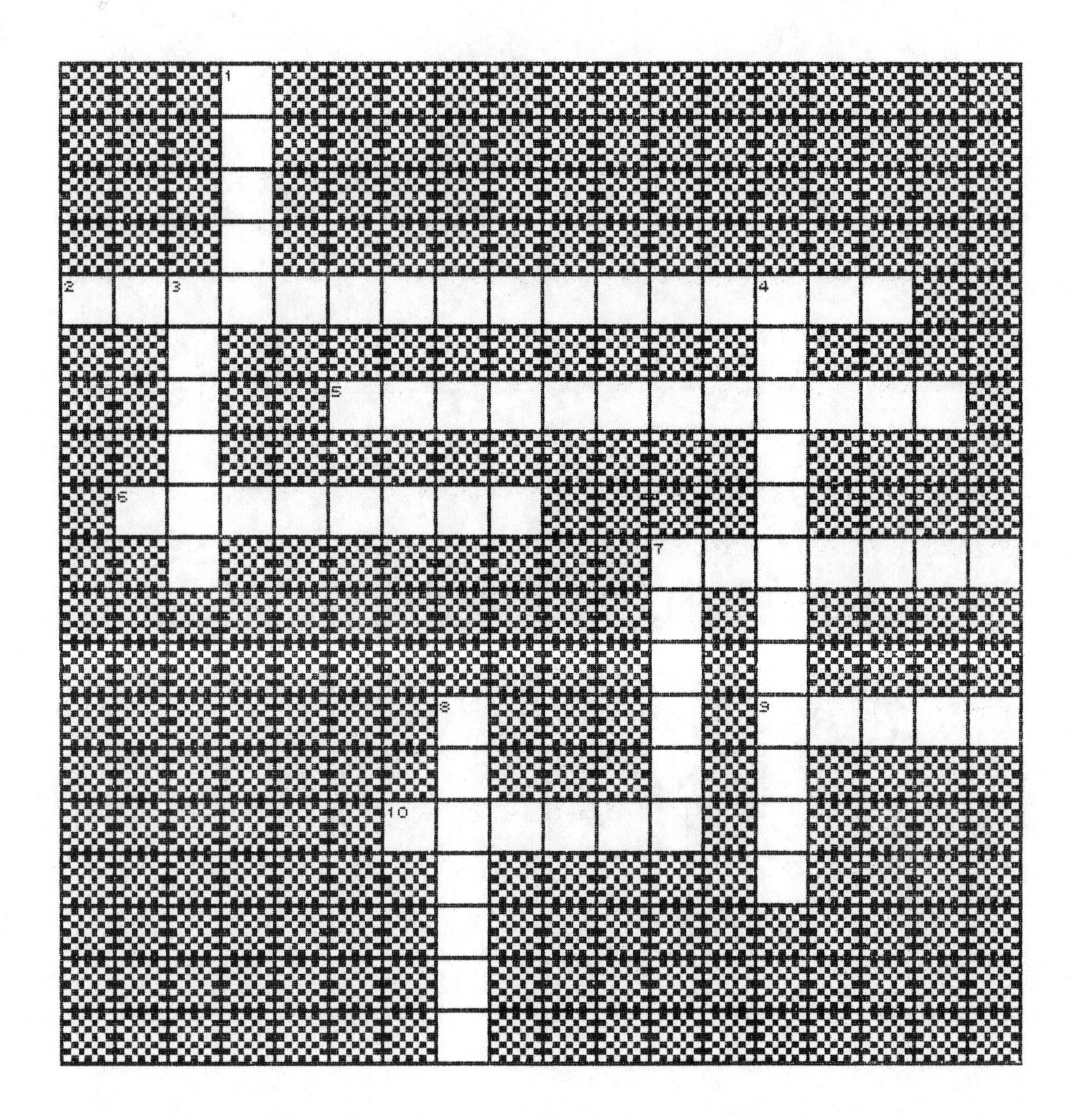

ACROSS

2. The sac that contains a mature ovum
5. Male sex hormone
6. Female erectile tissue
7. Sex cells
9. Male reproductive fluid
10. Menstrual period

DOWN

1. External genitalia
3. Colored area around nipple
4. Surgical removal of foreskin
7. Essential organs of reproduction
8. Foreskin

MALE REPRODUCTIVE ORGANS

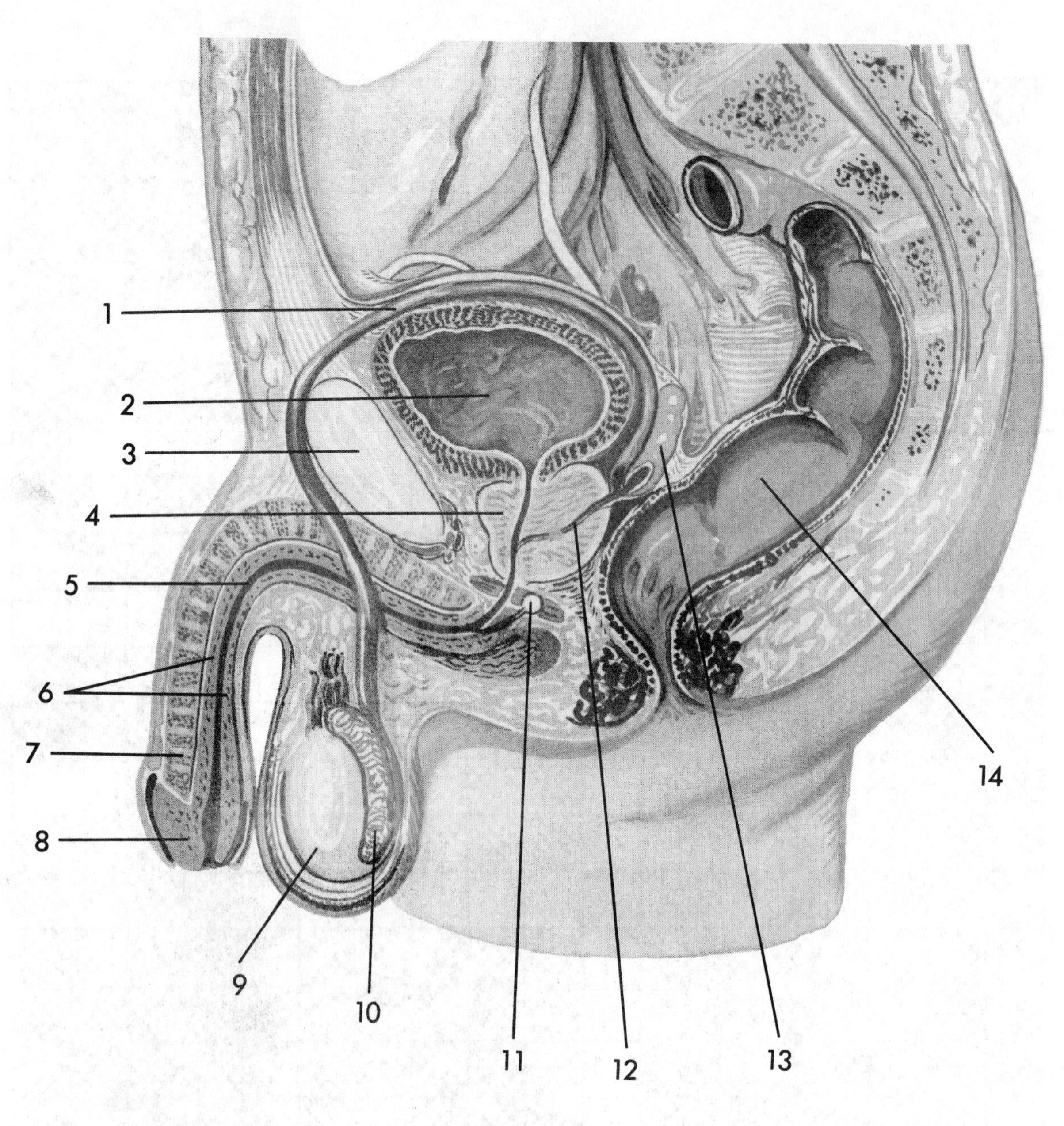

1. ______________________
2. ______________________
3. ______________________
4. ______________________
5. ______________________
6. ______________________
7. ______________________
8. ______________________
9. ______________________
10. ______________________
11. ______________________
12. ______________________
13. ______________________
14. ______________________

TUBULES OF TESTIS AND EPIDIDYMIS

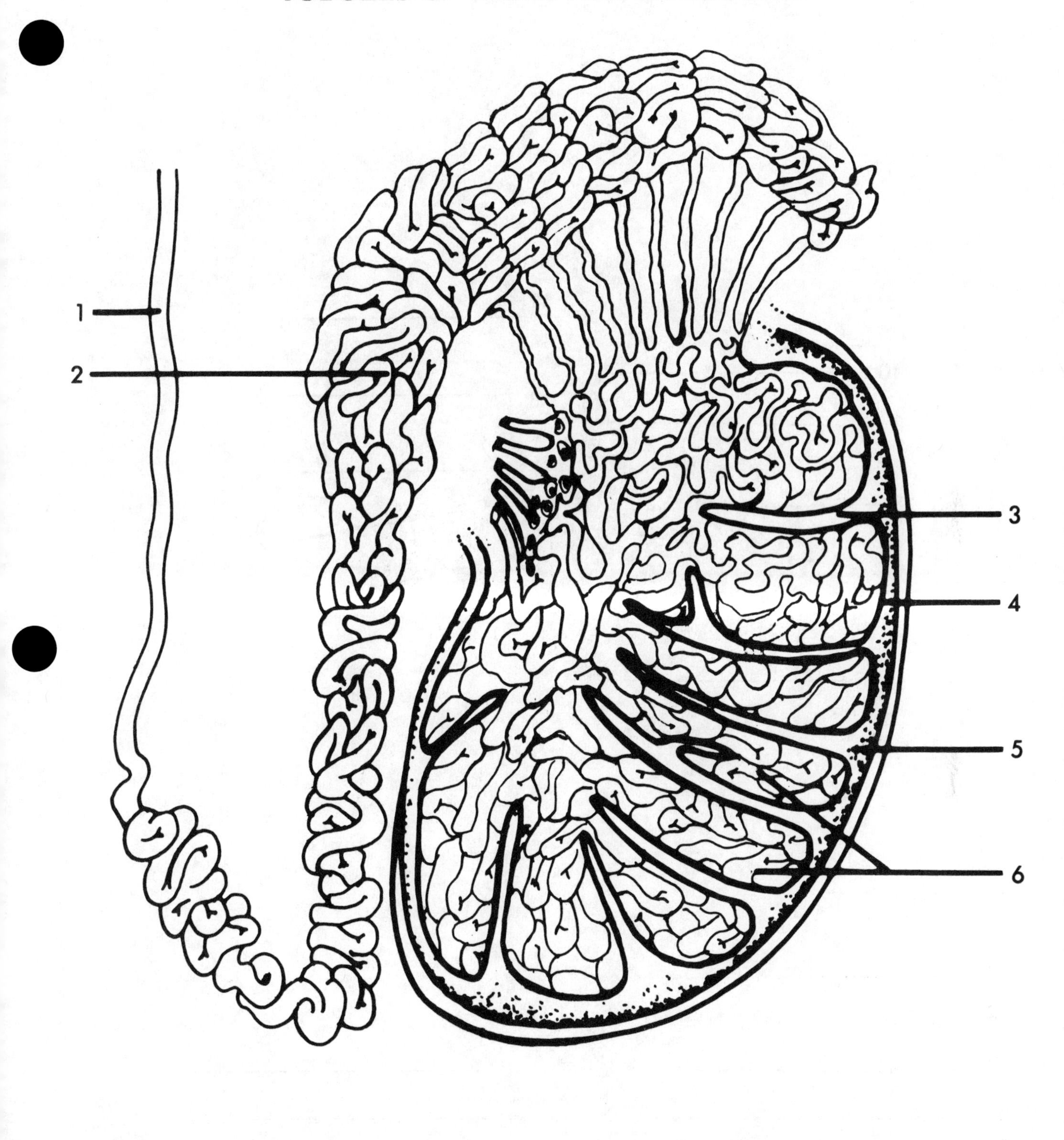

1. ______________________
2. ______________________
3. ______________________
4. ______________________
5. ______________________
6. ______________________

VULVA

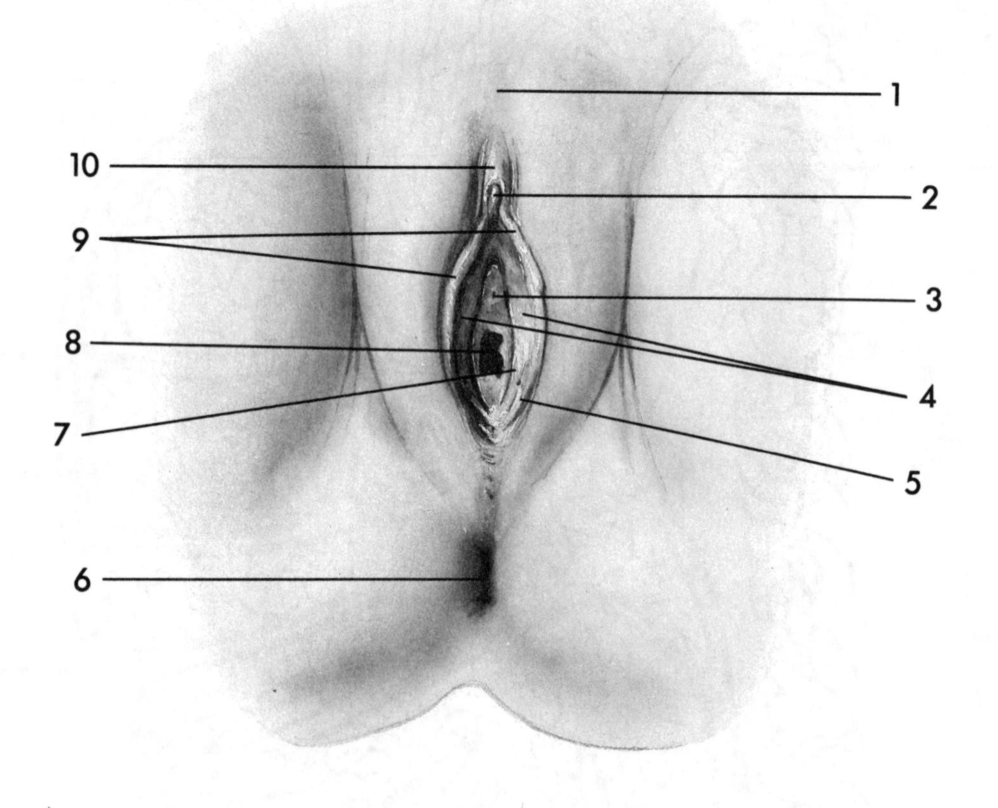

1. ______________________

2. ______________________

3. ______________________

4. ______________________

5. ______________________

6. ______________________

7. ______________________

8. ______________________

9. ______________________

10. ______________________

BREAST

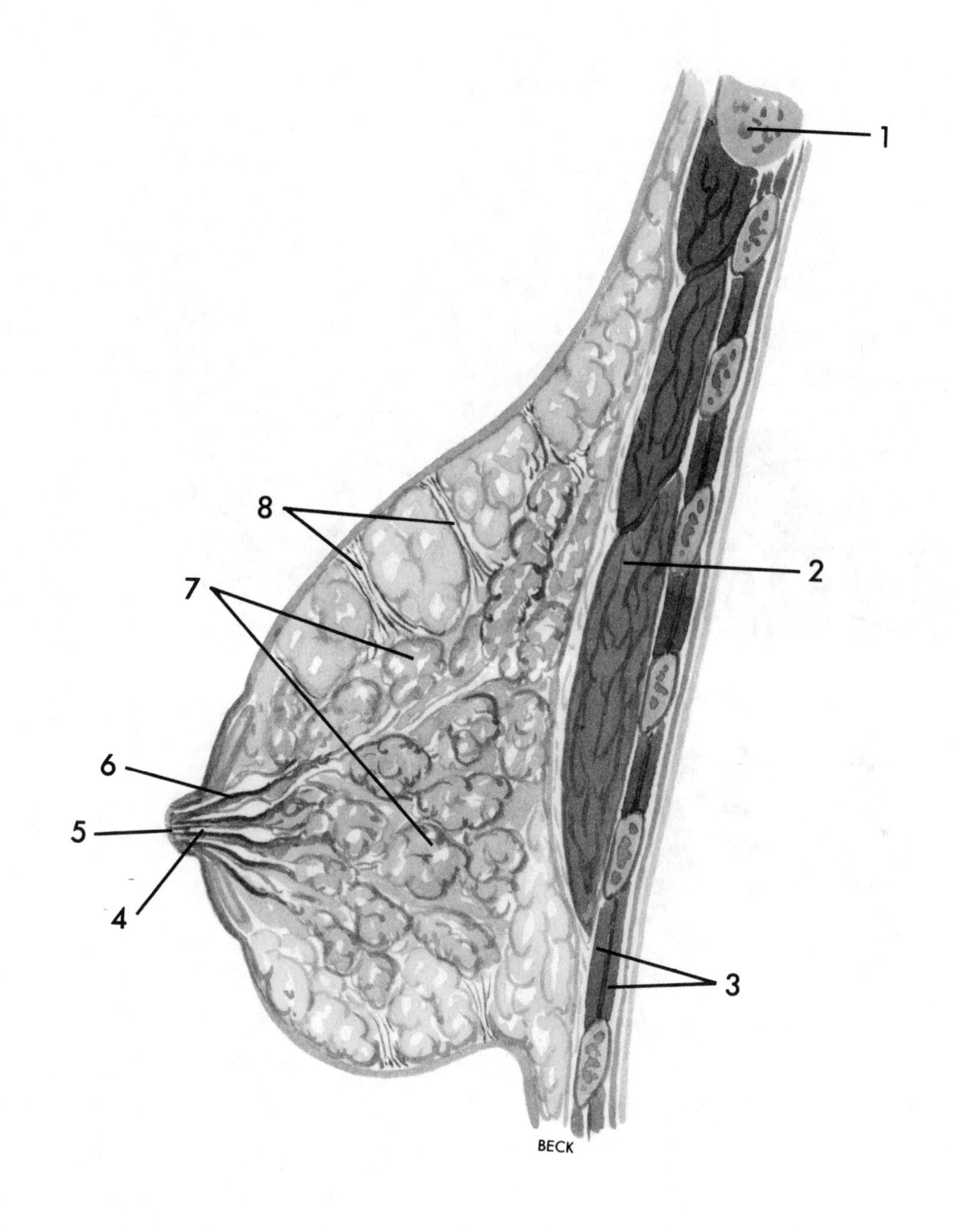

1. ______________________
2. ______________________
3. ______________________
4. ______________________
5. ______________________
6. ______________________
7. ______________________
8. ______________________

FEMALE PELVIS

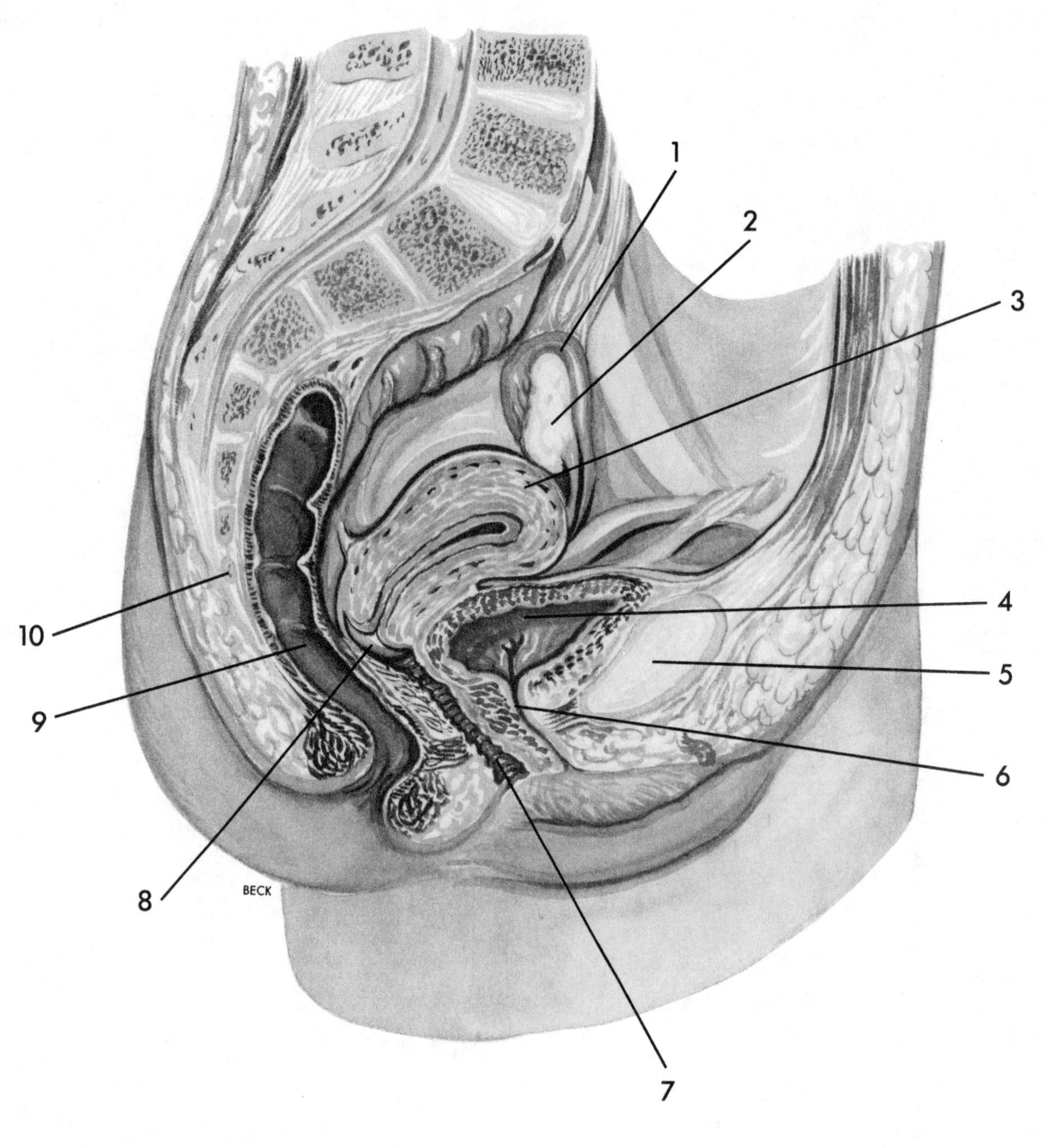

1. ______________________
2. ______________________
3. ______________________
4. ______________________
5. ______________________
6. ______________________
7. ______________________
8. ______________________
9. ______________________
10. ______________________

UTERUS AND ADJACENT STRUCTURES

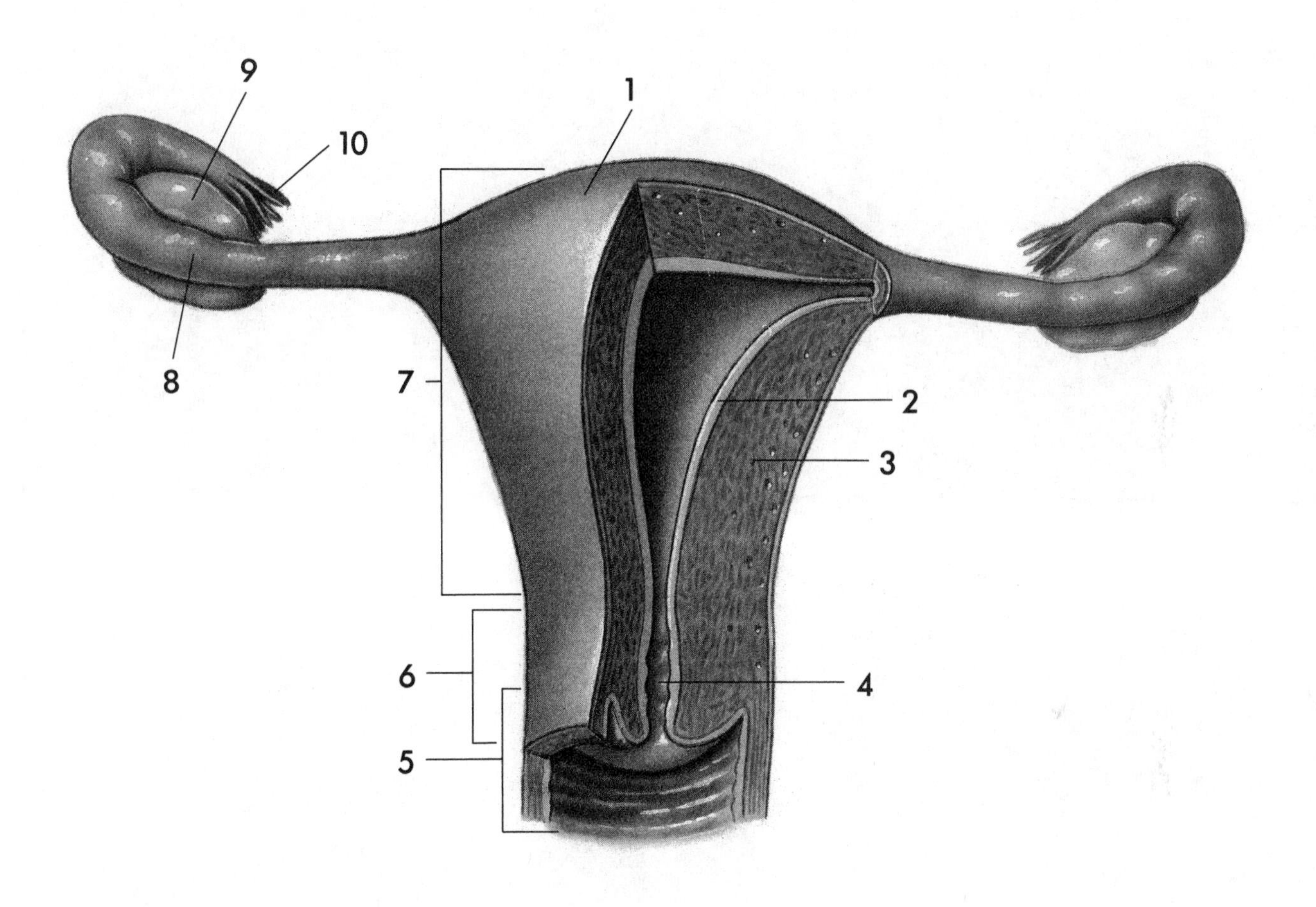

1. ______________________
2. ______________________
3. ______________________
4. ______________________
5. ______________________
6. ______________________
7. ______________________
8. ______________________
9. ______________________
10. ______________________

CHAPTER 20

Growth and Development

Millions of fragile microscopic sperm swim against numerous obstacles to reach the ova and create a new life. At birth, the newborn will fill his lungs with air and cry lustily, signaling to the world that he is ready to begin the cycle of life. This cycle will be marked by ongoing changes, periodic physical growth, and continuous development.

This chapter reviews the more significant events that occur in the normal growth and development of an individual from conception to death. Realizing that each individual is unique, we nonetheless can discover amid all the complexities of humanity some constants that are understandable and predictable.

A knowledge of human growth and development is essential in understanding the commonalities that influence individuals as they pass through the cycle of life.

TOPICS FOR REVIEW

Your review of this chapter should include an understanding of the concept of development as a biological process. You should familiarize yourself with the major developmental changes from conception through older adulthood. Your study should conclude with a review of the effects of aging on the body systems.

PRENATAL PERIOD

Fill in the blanks.

The prenatal stage of development begins at the time of (1) ___________ and continues until (2) ___________. The science of the development of an individual before birth is called (3) ________________. Fertilization takes place in the outer third of the (4) ________________. The fertilized ovum or (5) ________________ begins to divide and in approximately 3 days forms a solid mass called a (6) ______________. By the time it enters the uterus, it is a hollow ball of cells called a (7) ________________. As it continues to develop, it forms a structure with two cavities. The (8) ______________ will become a fluid-filled sac for the embryo. The (9) ______________ will develop into an important fetal membrane in the (10) ________________.

Choose the correct term and write its letter in the space next to the appropriate definition below.

a. Laparoscope
b. Gestation
c. Antenatal
d. Histogenesis
e. Quickening
f. Endoderm
g. In vitro
h. Parturition
i. Embryonic phase
j. Ultrasonogram

_____ 11. "Within a glass"
_____ 12. Inside germ layer
_____ 13. Before birth
_____ 14. Length of pregnancy
_____ 15. Optical viewing tube
_____ 16. Process of birth
_____ 17. First fetal movement
_____ 18. Study of how the primary germ layers develop into many different kinds of tissues
_____ 19. Fertilization until the end of the eighth week of gestation
_____ 20. Monitors progress of developing fetus

▸ If you have had difficulty with this section, review pages 402-412. ◂

POSTNATAL PERIOD

Circle the correct choice.

21. During the postnatal period:

    a. The head becomes proportionately smaller
    b. The relationship of the face to skull is reduced
    c. The legs become proportionately longer
    d. The trunk becomes proportionately shorter
    e. All of the above take place during the postnatal period

22. The period of infancy starts at birth and lasts about:

    a. 4 weeks
    b. 4 months
    c. 10 weeks
    d. 12 months
    e. 18 months

23. The lumbar curvature of the spine appears ____________ months after birth.

    a. 1-10
    b. 5-8
    c. 8-12
    d. 11-15
    e. 12-18

24. During the first 4 months the birth weight will:

    a. Double
    b. Triple
    c. Quadruple
    d. None of the above is correct

25. At the end of the first year the weight of the baby will have:

a. Doubled
b. Tripled
c. Quadrupled
d. None of the above is correct

26. The infant is capable of following a moving object with its eyes at:

a. 2 days
b. 2 weeks
c. 2 months
d. 4 months
e. 10 months

27. The infant can lift its head and raise its chest at:

a. 2 months
b. 3 months
c. 4 months
d. 10 months

28. The infant can crawl at:

a. 2 months
b. 3 months
c. 4 months
d. 10 months
e. 12 months

29. The infant can stand alone at:

a. 2 months
b. 3 months
c. 4 months
d. 10 months
e. 12 months

30. The permanent teeth, with the exception of the third molar, have all erupted by age ____________ years.

a. 6
b. 8
c. 12
d. 14
e. None of the above is correct

31. Puberty starts at age ____________ years in boys:

a. 10-13
b. 12-14
c. 14-16
d. None of the above is correct

32. Most girls begin breast development at about age:

a. 8
b. 9
c. 10
d. 11
e. 12

33. The growth spurt is generally complete by age __________ in males:

a. 14
b. 15
c. 16
d. 18

34. An average age at which girls begin to menstruate is _________ years.

a. 10-12
b. 11-12
c. 12-13
d. 13-14
e. 14-15

35. The first sign of puberty in boys is:

a. Facial hair
b. Increased muscle mass
c. Pubic hair
d. Deepening of the voice
e. Increased testicular enlargement

Write the letter of the correct word in the blank next to the appropriate definition.

a. Neonatology
b. Neonatal
c. Adolescence
d. Deciduous
e. Puberty
f. Postnatal
g. Infancy
h. Childhood
i. Senescence

_____36. Begins at birth and lasts until death
_____37. Concerned with the diagnosis and treatment of disorders of the newborn
_____38. Teenage years
_____39. From the end of infancy to puberty
_____40. Baby teeth
_____41. First 4 weeks of infancy
_____42. Secondary sexual characteristics occur
_____43. Begins at birth and lasts about 18 months
_____44. Old age

▸ If you have had difficulty with this section, review pages 412-415. ◂

EFFECTS OF AGING

Fill in the blanks.

45. Old bones develop indistinct and shaggy margins with spurs; a process called _________________.

46. A degenerative joint disease common in the aged is _________________.

47. The number of _____________ units in the kidney decreases by almost 50% between the ages of 30 and 75.

48. In old age, respiratory efficiency decreases, and a condition known as _____________ _____________ results.

49. Fatty deposits accumulate in blood vessels as we age, and the result is _______________ which narrows the passageway for the flow of blood.

50. Hardening of the arteries or ______________ occurs during the aging process.

51. Another term for high blood pressure is ________________.

52. Hardening of the lens is ________________.

53. If the lens becomes cloudy and impairs vision, it is called a ________________.

54. ________________ causes an increase in the pressure within the eyeball and may result in blindness.

▸ If you have had difficulty with this section, review pages 415-418. ◂

Unscramble the words.

55. ANNFCYI

56. NAALTTSOP

57. OGSSNEGRAONEI

58. GTEYZO

59. HDOOLHCID

Take the circled letters, unscramble them, and fill in the statement.

The secret to Farmer Brown's prize pumpkin crop.

60.

APPLYING WHAT YOU KNOW

61. Heather's mother told the pediatrician during her 1-year visit that she had tripled her birth weight, was crawling actively, and could stand alone. Is Heather's development normal, retarded, or advanced?

62. Clarke is 70 years old. She has always enjoyed food and has had a hearty appetite. Lately, however, she has complained that food "just doesn't taste as good anymore." What might be a possible explanation?

63. Mr. Ruiz, age 68, has noticed hearing problems, but only under certain circumstances. He has difficulty with certain tones, especially high or low tones, but has no problem with everyday conversation. What might be a possible explanation?

GROWTH/DEVELOPMENT

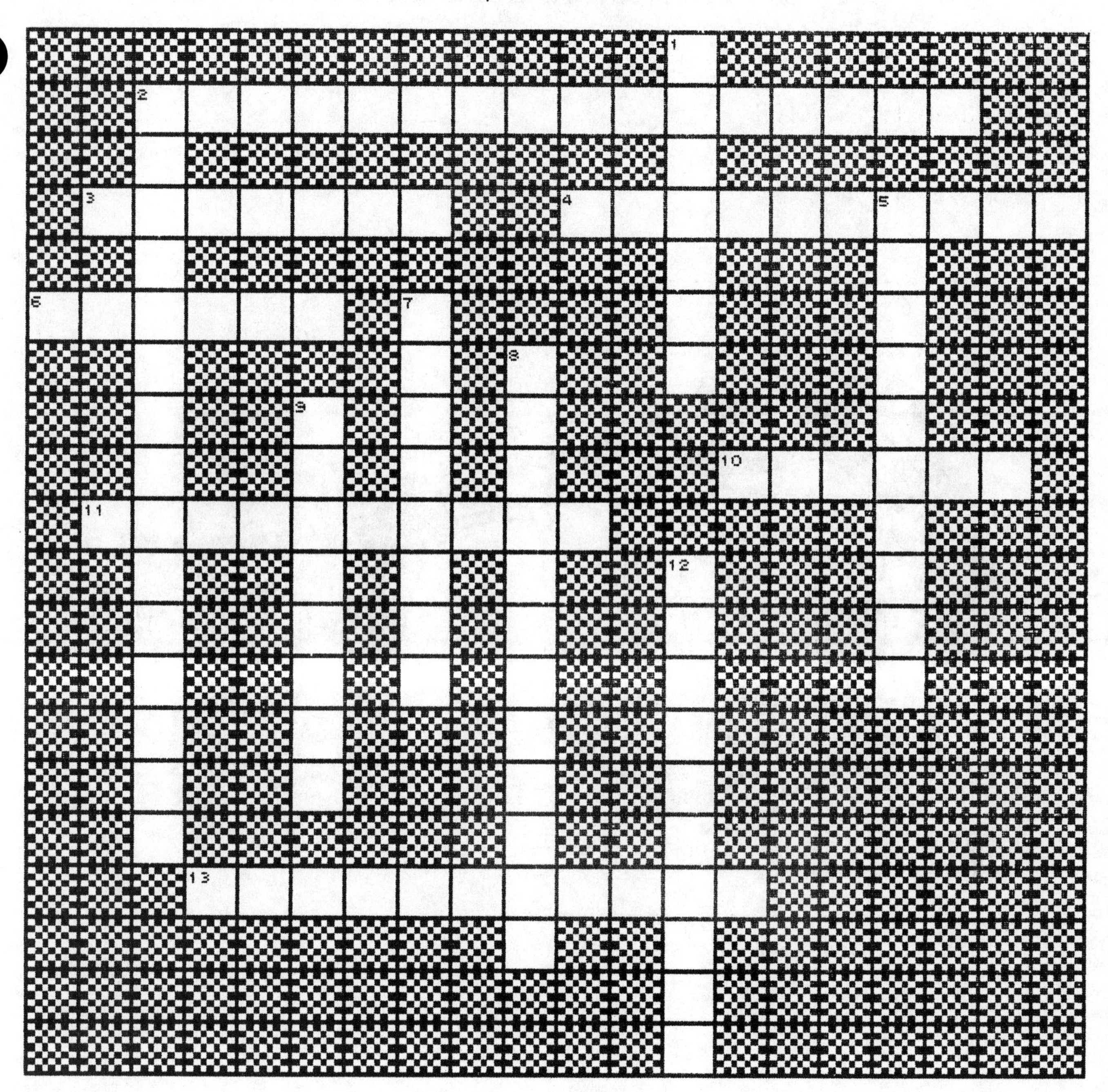

ACROSS

2. Hardening of the arteries
3. Will develop into a fetal membrane in the placenta
4. Old age
6. Name of zygote after 3 days
10. Fertilized ovum
11. Name of zygote after implantation
13. Process of birth

DOWN

1. First 4 weeks of infancy
2. Fatty deposit buildup on walls of arteries
5. Science of the development of the individual before birth
7. Eye disease marked by increased pressure in the eyeball
8. Study of how germ layers develop into tissues
9. Cloudy lens
12. Hardening of the lens

FERTILIZATION AND IMPLANTATION

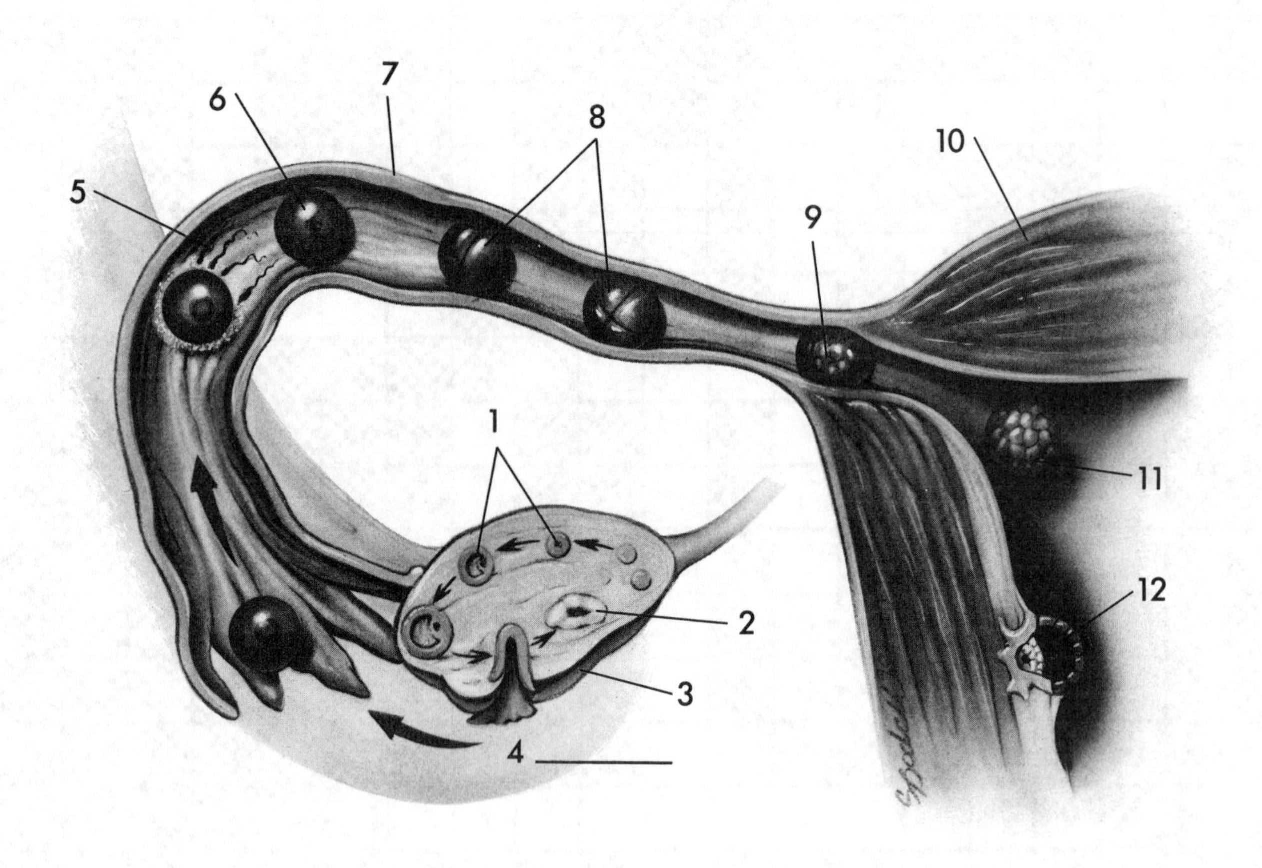

1.	______	7.	______
2.	______	8.	______
3.	______	9.	______
4.	______	10.	______
5.	______	11.	______
6.	______	12.	______

ANSWER KEY

ANSWERS TO EXERCISES

CHAPTER 1
AN INTRODUCTION TO THE STRUCTURE AND FUNCTION OF THE BODY

Matching

Group A

1. D, p. 2
2. E, p. 2
3. A, p. 2
4. C, p. 2
5. B, p. 2

Group B

6. C, p. 6
7. E, p. 6
8. A, p. 4
9. B, p. 6
10. D, p. 4

CROSSWORD

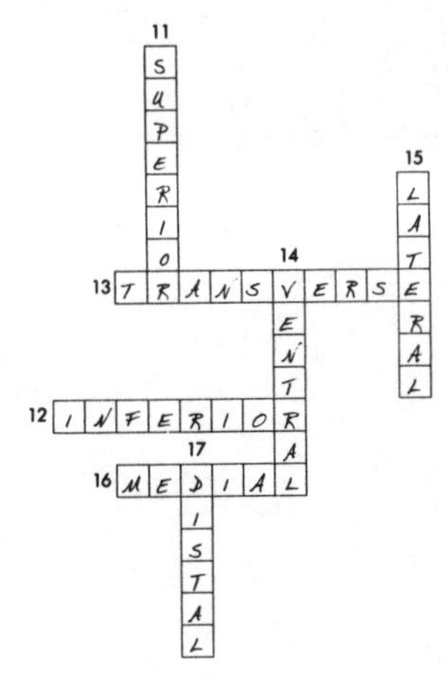

11. Superior
12. Inferior
13. Transverse
14. Ventral
15. Lateral
16. Medial
17. Distal

Did you notice that the answers were arranged as they appear on the human body?

Circle the correct answer

18. Inferior, p. 5
19. Anterior, p. 6
20. Lateral, p. 6
21. Proximal, p. 6
22. Superficial, p. 6
23. Equal, p. 9
24. Anterior and posterior, p. 9
25. Upper and lower, p. 9
26. Right upper quadrant, p. 8
27. Forward, p. 9

Circle the one that does not belong

28. Extremities (all others are part of the axial portions)
29. Cephalic (all others are part of the arm)
30. Plantar (all others are part of the face)
31. Eyes closed (all others are part of the anatomical position)
32. Plantar (all others are part of the trunk)
33. Carpal (all others are part of the leg or foot)
34. Thoracic (all others are located on the posterior sections of the body)

Fill in the blanks

35. Survival, p. 12
36. Internal environment, p. 12
37. Rise to toxic levels, p. 13
38. Developmental processes, p. 13
39. Aging processes, p. 13
40. Negative, positive, p. 13
41. Stabilizing, p. 13
42. Stimulatory, p. 13

APPLYING WHAT YOU KNOW

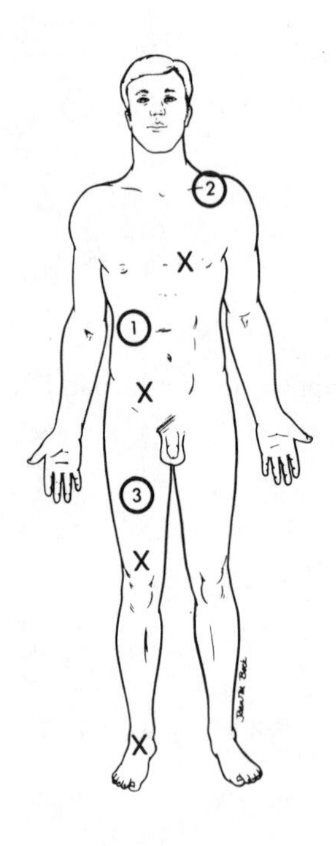

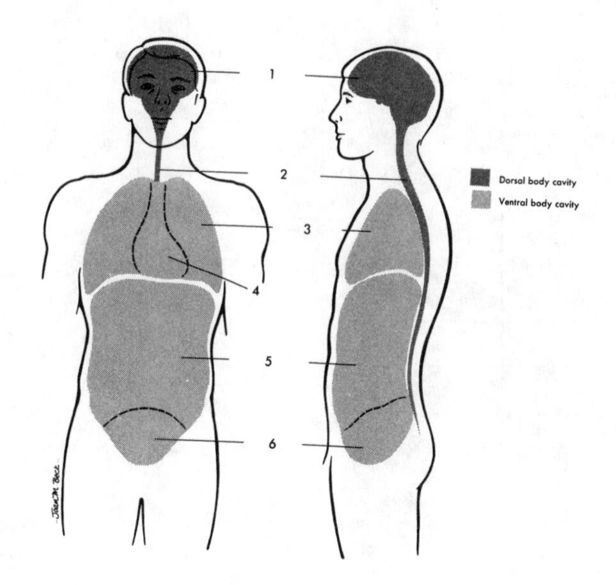

Dorsal and Ventral Body Cavities

1. Cranial cavity
2. Spinal cavity
3. Thoracic cavity
4. Mediastinum
5. Abdominal cavity
6. Pelvic cavity

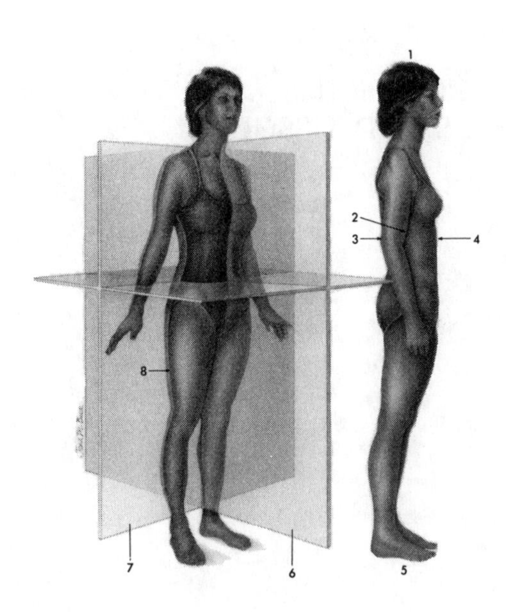

Directions and Planes of Body

1. Superior
2. Proximal
3. Posterior (Dorsal)
4. Anterior (Ventral)
5. Inferior
6. Sagittal plane
7. Frontal plane
8. Lateral

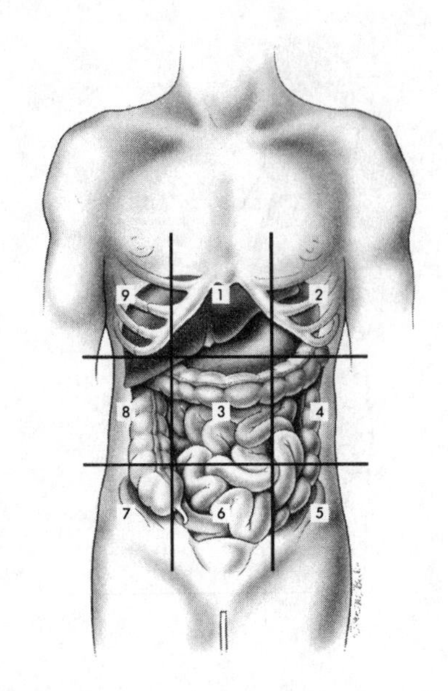

Regions of the Abdomen

1. Epigastric region
2. Left hypochondriac region
3. Umbilical region
4. Left lumbar region
5. Left iliac (inguinal) region
6. Hypogastric region
7. Right iliac (inguinal) region
8. Right lumbar region
9. Right hypochondriac region

CHAPTER 2
CELLS AND TISSUES

Matching

Group A

1. C, p. 18
2. E, p. 18
3. A, p. 18
4. B, p. 23
5. D, p. 22

Group B

6. D, p. 22
7. E, p. 22
8. A, p. 22
9. B, p. 22
10. C, p. 22

Fill in the blanks

11. Metric System, p. 19
12. Tissue typing, p. 20
13. Cilia, p. 22
14. Endoplasmic reticulum, p. 20
15. Ribosomes, p. 22
16. Mitochondria, p. 22
17. Lysosomes, p. 22
18. Golgi apparatus, p. 22
19. Centrioles, p. 22
20. Chromatin granules, p. 23

Multiple choice

21. A, p. 24
22. D, p. 24
23. B, p. 24
24. D, p. 24
25. C, p. 24
26. A, p. 24
27. B, p. 24
28. C, p. 26
29. C, p. 26
30. A, p. 26
31. B, p. 25
32. A, p. 25

Circle the one that does not belong

33. Uracil (RNA contains the base uracil, not DNA)
34. RNA (the others are complementary base pairings of DNA)
35. Anaphase (the others refer to genes and heredity)
36. Thymine (the others refer to RNA)
37. Mitosis (the others refer to DNA before mitosis)
38. Prophase (the others refer to anaphase)
39. Prophase (the others refer to interphase)
40. Metaphase (the others refer to telophase)
41. Gene (the others refer to stages of cell division)

42. **Fill in the missing area**

TISSUE	LOCATION	FUNCTION
1.	1a.	1a. Absorption by diffusion of respiratory gases between alveolar air and blood
	1b.	1b. Absorption by diffusion, filtration and osmosis
2.	2a. Surface of lining of mouth and esophagus 2b. Surface of skin	2.
3.	3. Surface layer of lining of stomach, intestines, and parts of respiratory tract	3.
4. Stratified transitional	4.	4.
5.	5. Surface of lining of trachea	5.
1.	1. Between other tissues and organs	1.
2. Adipose	2.	2.
3.	3.	3. Flexible but strong connection
4.	4. Skeleton	4.
5.	5. Part of nasal septum, larynx, rings in trachea and bronchi, disks between vertebrae, external ear	5.
6.	6.	6. Transportation
7. Hemopoietic tissue	7.	7.
1.	1. Muscles that attach to bones, eyeball muscles, upper third of esophagus	1.
2. Cardiac	2.	2.

3.	3. Walls of digestive, respiratory, and genitourinary tracts; walls of blood and large lymphatic vessels; ducts of glands; intrinsic eye muscles; arrector muscles of hair	3.
1.	1. Brain and spinal cord, nerves	1.

APPLYING WHAT YOU KNOW

43.

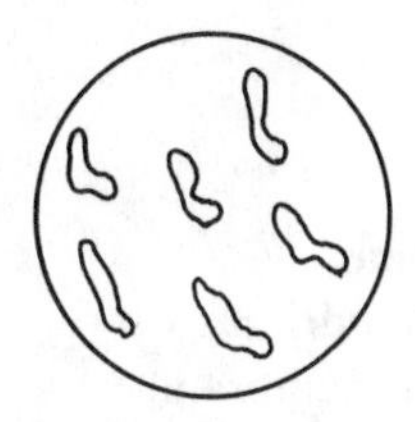

44. Diffusion
45. Simple squamous epithelium
46. Merrily may have exceeded the 15-18% desirable body fat composition. Fitness depends more on the percentage and ratio of specific tissue types than the overall amount of tissue present.

CROSSWORD

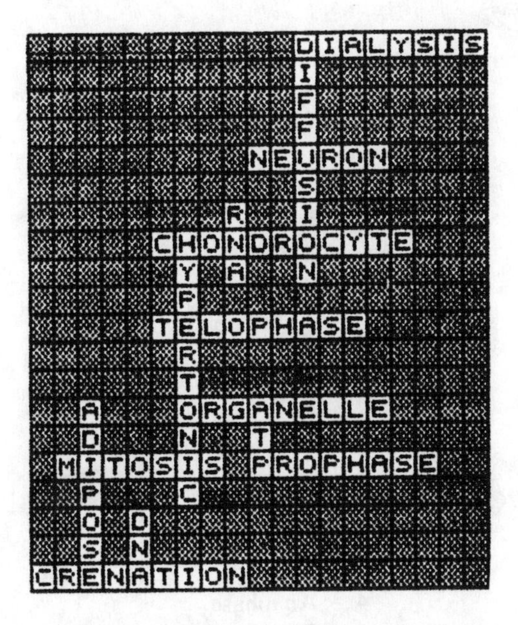

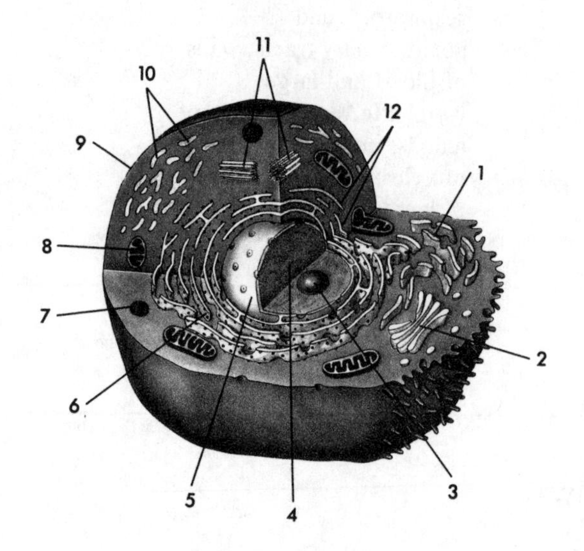

Cell Structure

1. Smooth endoplasmic reticulum
2. Golgi apparatus
3. Nucleolus
4. Nucleus
5. Nuclear membrane
6. Rough endoplasmic reticulum
7. Lysosome
8. Mitochondrion
9. Plasma membrane
10. Smooth endoplasmic reticulum
11. Centrioles
12. Ribosomes

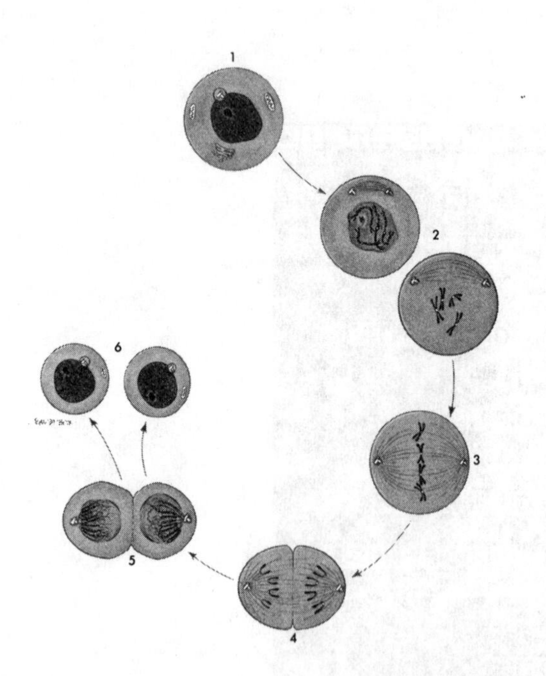

Mitosis

1. Interphase
2. Prophase
3. Metaphase
4. Anaphase
5. Telophase
6. Daughter cells (interphase)

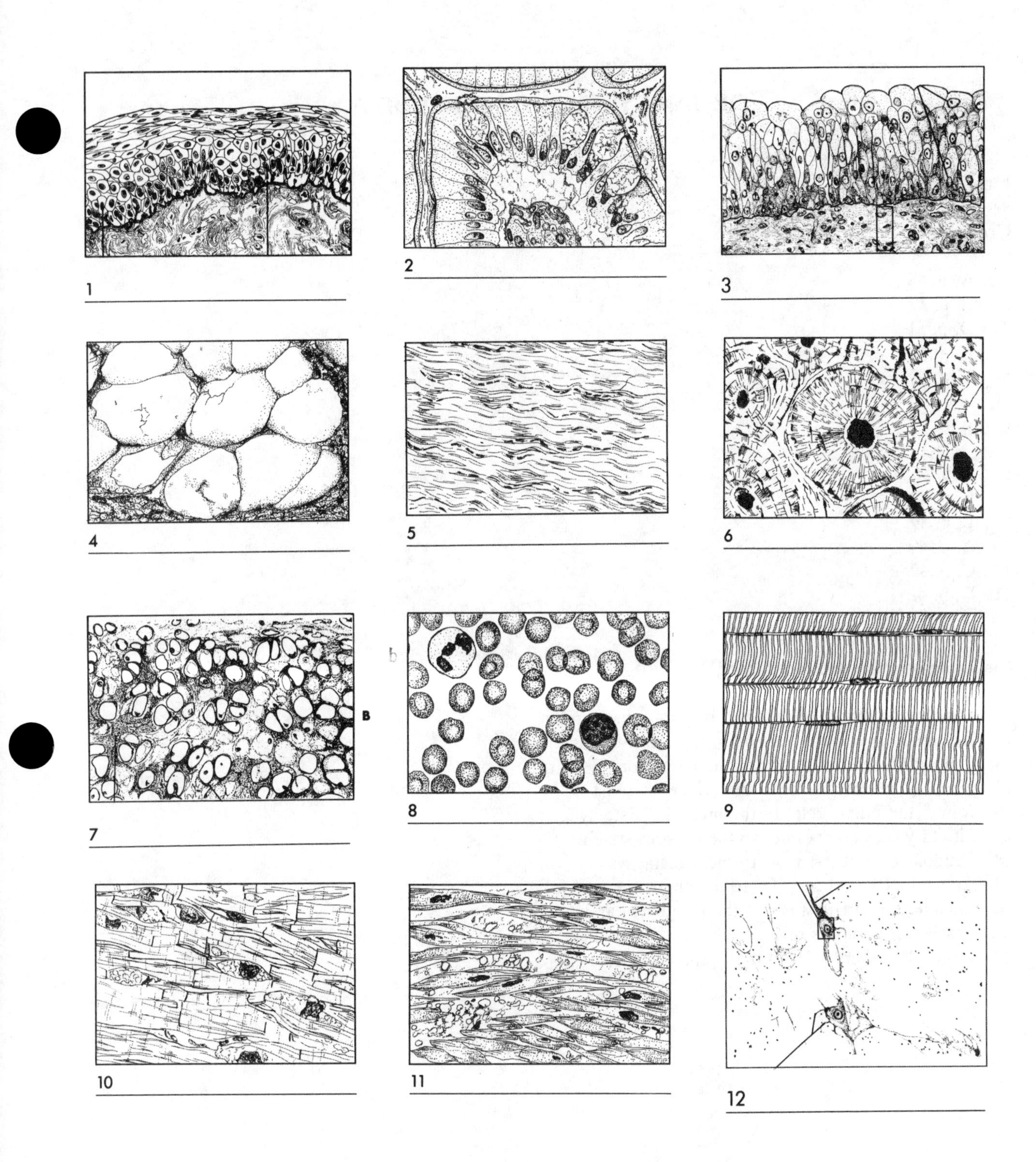

Tissues

1. Stratified squamous epithelium
2. Simple columnar epithelium
3. Stratified transitional epithelium
4. Adipose tissue
5. Dense fibrous connective tissue
6. Bone tissue
7. Cartilage
8. Blood
9. Skeletal muscle
10. Cardiac muscle
11. Smooth muscle
12. Nervous tissue

CHAPTER 3
ORGAN SYSTEMS OF THE BODY

Matching

Group A

1. A, p. 48
2. E, p. 49
3. D, p. 51
4. B, p. 49
5. C, p. 53

Group B

6. F, p. 55
7. E, p. 49
8. B, p. 49
9. A, p. 49
10. C, p. 49
11. D, p. 49

Circle the one that does not belong

12. Mouth (the others refer to the respiratory system)
13. Rectum (the others refer to the reproductive system)
14. Pancreas (the others refer to the circulatory system)
15. Pineal (the others refer to the urinary system)
16. Joints (the others refer to the muscular system)
17. Pituitary (the others refer to the nervous system)
18. Tendons (the others refer to the skeletal system)
19. Appendix (the others refer to the endocrine system)
20. Thymus (the others refer to the integumentary system)
21. Trachea (the others refer to the digestive system)
22. Liver (the others refer to the lymphatic system)

Fill in the missing area

SYSTEM	ORGAN	FUNCTIONS
23.		Protection, regulation of body temperature, synthesis of chemicals and hormones, serves as a sense organ
24.	Bones, joints	
25.		Movement, maintains body posture, produces heat
26. Nervous		
27.	Pituitary, thymus, pineal, adrenal, hypothalamus, thyroid, pancreas, parathyroid, ovaries, testes	
28.		Transportation, immunity
29.	Lymph nodes, lymph vessels, thymus, spleen, tonsils	
30. Urinary		
31.	Mouth, pharynx, esophagus, stomach, small and large intestine, rectum, anal canal, teeth, salivary glands, tongue, liver, gallbladder, pancreas, appendix	
32. Respiratory		
33.	a. Gonads - testes and ovaries b. Accessory glands (p. 49) Supporting structures (p. 49)	

Unscramble the words

34. Heart
35. Pineal
36. Nerve
37. Esophagus
38. Nervous

APPLYING WHAT YOU KNOW

39. (a) Endocrinology (endocrine system)
    (b) Gynecology (reproductive system)

40. The skin protects the underlying tissue against invasion by harmful bacteria. With a large percentage of Brian's skin destroyed, he was vulnerable to bacteria, and so he was placed in the cleanest environment possible - isolation. Jenny is required to wear special attire so that the risk of a visitor bringing bacteria to the patient is reduced.

CROSSWORD

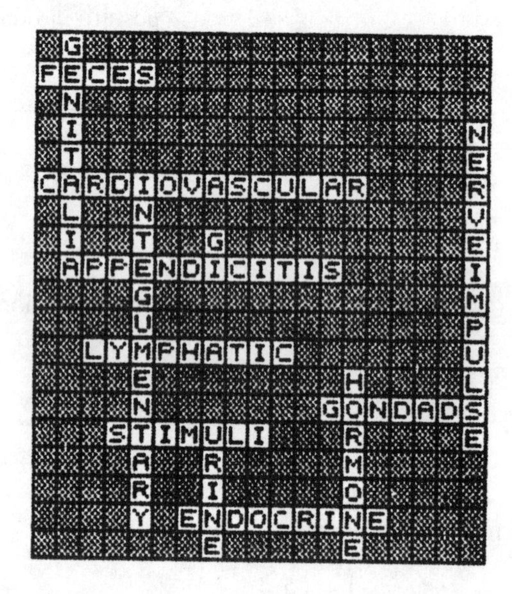

CHAPTER 4
THE INTEGUMENTARY SYSTEM AND BODY MEMBRANES

Select the best answer

1. B, p. 64
2. D, p. 66
3. C, p. 66
4. A, p. 64
5. B, p. 64
6. D, p. 66
7. C, p. 66
8. C, p. 66

Matching

Group A

9. D, p. 66
10. A, p. 66
11. B, p. 66
12. C, p. 69
13. E, p. 66

Group B

14. A, p. 67
15. D, p. 68
16. E, p. 68
17. C, p. 69
18. B, p. 68

Select the best answer

19. A, p. 67
20. B, p. 69
21. B, p. 69
22. A, p. 68
23. A, p. 67
24. B, p. 66
25. B, p. 70
26. B, p. 73
27. B, p. 72
28. A, p. 68 (Fig. 4-3)

Fill in the blanks

29. Protection, temperature regulation, and sense organ activity, p. 73
30. Melanin, p. 73
31. Lanugo, p. 70
32. Hair papillae, p. 70
33. Protein, p. 70
34. Arrector pili, p. 70
35. Light touch, p. 71
36. Eccrine, p. 72
37. Apocrine, p. 72
38. Sebum, p. 73

Circle the correct answer

39. Will not, p. 75
40. Will, p. 75
41. Will not, p. 75
42. 11, p. 75
43. Third, p. 75

Unscramble the words

44. Epidermis
45. Keratin
46. Hair
47. Lanugo
48. Dehydration
49. Third Degree

APPLYING WHAT YOU KNOW

50. 46%
51. Pleurisy

CROSSWORD

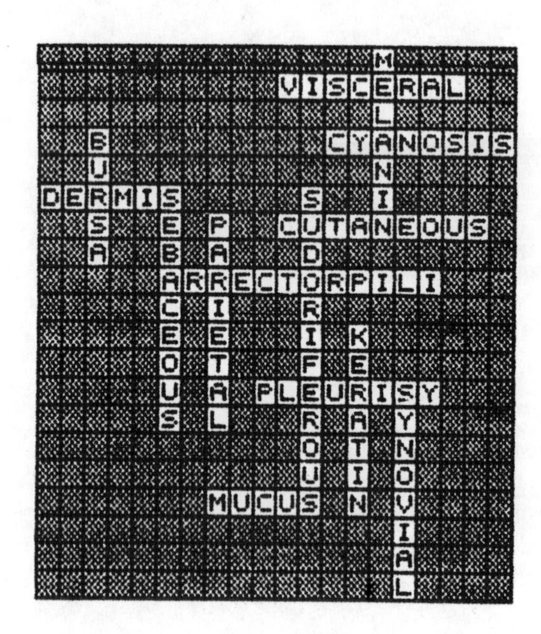

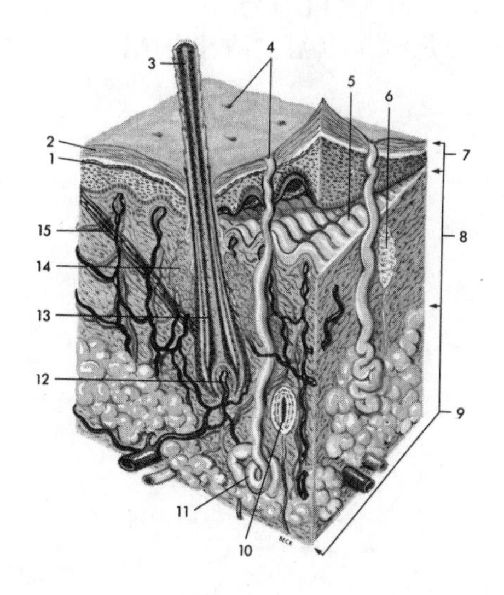

Longitudinal Section of the Skin

1. Pigment layer
2. Stratum corneum
3. Hair shaft
4. Openings of sweat ducts
5. Dermal papilla
6. Meissner's corpuscle
7. Epidermis
8. Dermis
9. Subcutaneous fatty tissue
10. Pacinian corpuscle
11. Sweat gland
12. Papilla of hair
13. Hair follicle
14. Sebaceous (oil) gland
15. Arrector pili muscle

Rule of Nines

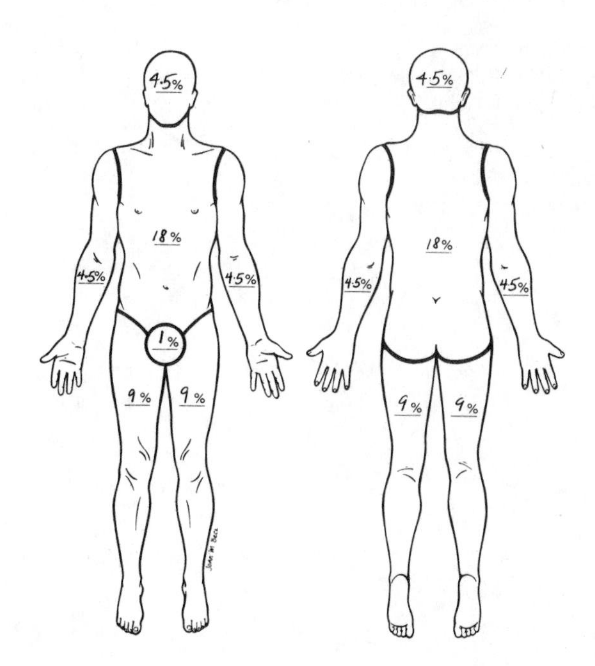

CHAPTER 5
THE SKELETAL SYSTEM

True or false

1. T
2. Epiphyses, not diaphyses, p. 80
3. Osteoblasts, not osteoclasts, p. 80
4. T
5. Increase, not decrease, p. 80
6. Juvenile, not adult, p. 80
7. Diaphysis, not articulation, p. 80
8. T
9. Ceases, not begins, p. 80
10. T

Matching

Group A

11. D, p. 82
12. B, p. 82
13. E, p. 80
14. A, p. 83
15. C, p. 83

Group B

16. D, p. 83
17. A, p. 83
18. E, p. 82
19. B, p. 83
20. C, p. 82

Fill in the blanks

21. 4, p. 83
22. Medullary cavity, p. 84
23. Articular cartilage, p. 85
24. Endosteum, p. 85
25. Hemopoiesis, p. 85

Multiple choice

26. A, p. 85
27. D, p. 88
28. A, p. 91
29. D, p. 93
30. C, p. 95

31. C, p. 96
32. D, p. 95
33. D, p. 96
34. A, p. 98
35. B, p. 88
36. A, p. 93
37. B, p. 95
38. B, p. 93
39. B, p. 95
40. C, p. 96
41. A, p. 96
42. D, p. 88
43. C, p. 90
44. C, p. 88

Circle the one that does not belong

45. Os coxae (all others refer to the spine)
46. Axial (all others refer to the appendicular skeleton)
47. Occipital (all others refer to a pair of sinuses)
48. Ribs (all others refer to the shoulder girdle)
49. Vomer (all others refer to the bones of the middle ear)
50. Ulna (all others refer to the os coxae bone)
51. Ethmoid (all others refer to the hand and wrist)
52. Nasal (all others refer to cranial bones)
53. Anvil (all others refer to the cervical vertebra)

Choose the correct answer

54. A, p. 99
55. B, p. 99
56. B, p. 84
57. A, p. 99
58. B, p. 99

Matching

59. C, p. 88
60. G, p. 92
61. J, L, M, and K, p. 99
62. N, p. 99
63. I, p. 95
64. A, p. 88
65. P, p. 99
66. D, B, p. 88
67. F, p. 88
68. H, Q, p. 95
69. O, T, p. 99
70. R, p. 88
71. S, E, p. 88

Choose the correct answer

72. Diarthroses, p. 102
73. Synarthrotic, p. 101
74. Diarthrotic, p. 102
75. Ligaments, p. 102
76. Articular cartilage, p. 102
77. Least movable, p. 103
78. Largest, p. 104
79. 2, p. 102
80. Mobility, p. 103
81. Pivot, p. 103

Unscramble the words

82. Vertebrae
83. Pubis
84. Scapula
85. Mandible
86. Phalanges
87. Pelvic girdle

APPLYING WHAT YOU KNOW

88. The bones are responsible for the majority of our blood cell formation. The disease condition of the bones might be inhibiting the production of blood cells for Mrs. Perine.

89. Epiphyseal cartilage is present only while a child is still growing. It becomes bone in adulthood. It is particularly vulnerable to fractures in childhood and preadolescence.

90. Osteoporosis

CROSSWORD

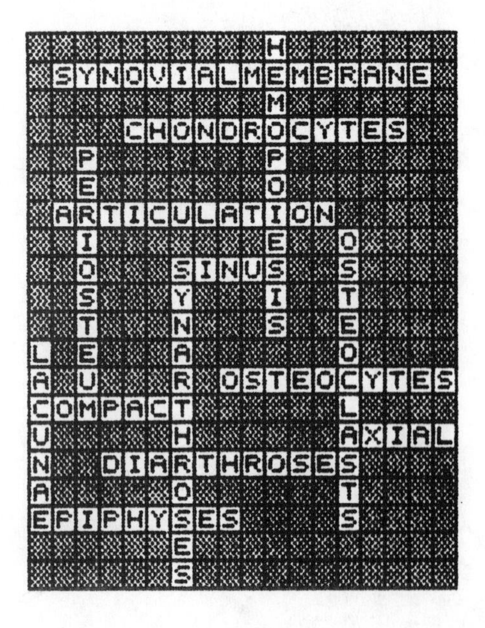

Long Bone

1. Articular cartilage
2. Spongy bone
3. Epiphyseal plate
4. Red marrow cavities
5. Compact bone
6. Medullary cavity
7. Yellow marrow
8. Periosteum

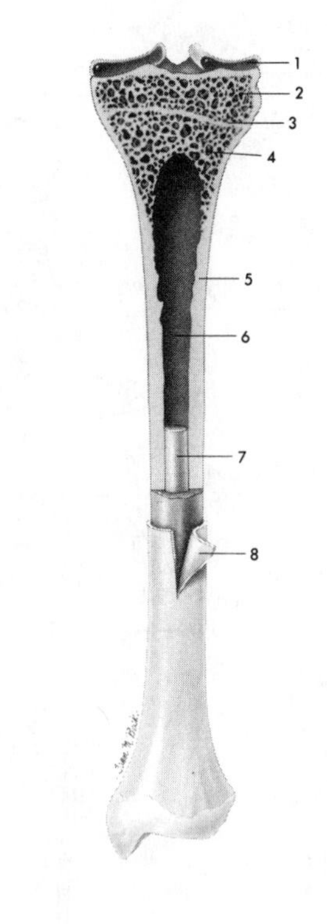

Anterior View of Skeleton

1. Orbit
2. Mandible
3. Sternum
4. Xiphoid process
5. Costal cartilage
6. Os coxae
7. Ilium
8. Pubis
9. Ischium
10. Frontal
11. Nasal
12. Maxilla
13. Clavicle
14. Ribs
15. Humerus
16. Vertebral column
17. Ulna
18. Radius
19. Sacrum
20. Coccyx
21. Carpals
22. Metacarpals
23. Phalanges
24. Femur
25. Patella
26. Tibia
27. Fibula
28. Tarsals
29. Metatarsals
30. Phalanges

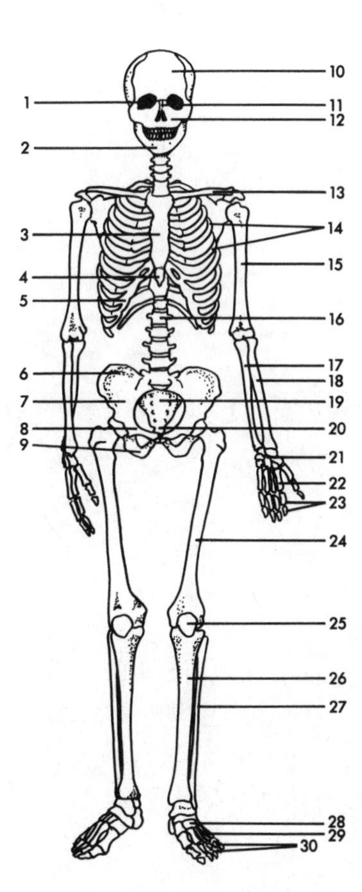

Posterior View of Skeleton

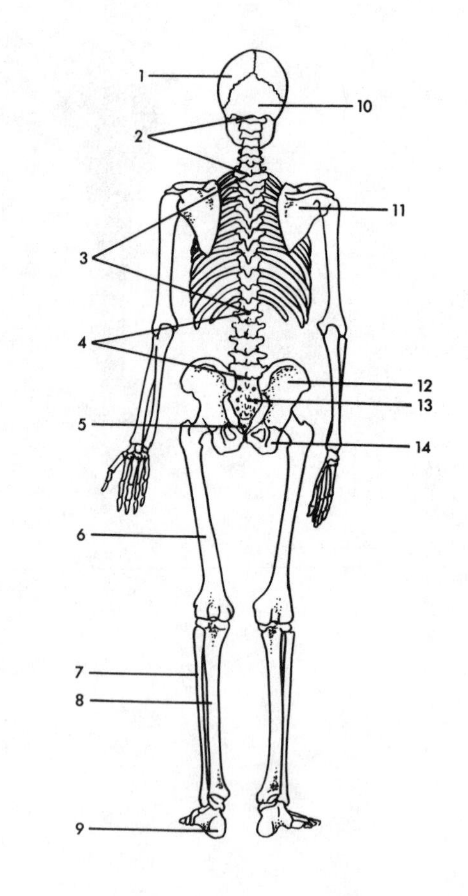

1. Parietal
2. Cervical vertebrae
3. Thoracic vertebrae
4. Lumbar vertebrae
5. Coccyx
6. Femur
7. Fibula
8. Tibia
9. Calcaneus
10. Occipital
11. Scapula
12. Ilium
13. Sacrum
14. Ischium

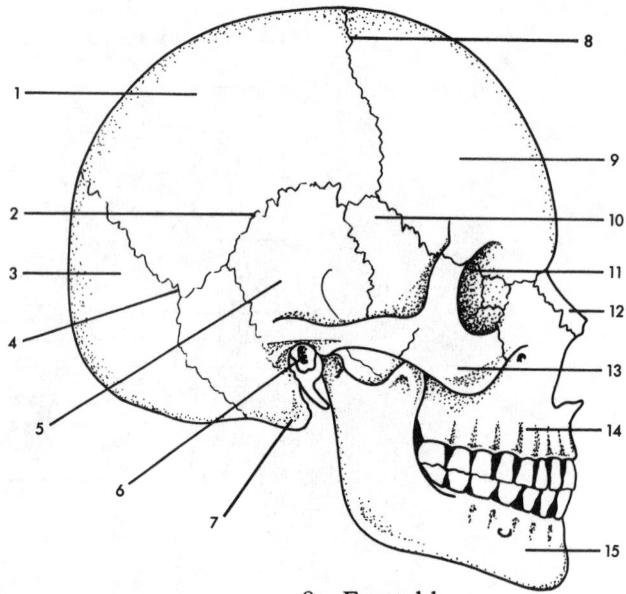

Skull - Right side

1. Parietal bone
2. Squamous suture
3. Occipital bone
4. Lambdoidal suture
5. Temporal bone
6. External auditory canal
7. Mastoid process
8. Coronal suture
9. Frontal bone
10. Sphenoid bone
11. Ethmoid bone
12. Nasal bone
13. Zygomatic bone
14. Maxilla
15. Mandible

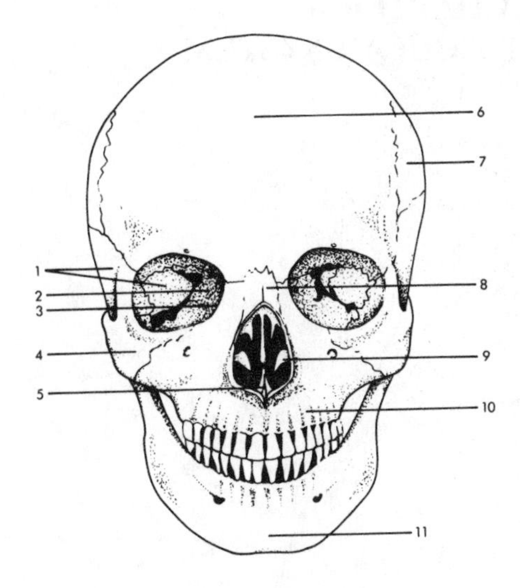

Skull - Front view

1. Sphenoid bone
2. Ethmoid bone
3. Lacrimal bone
4. Zygomatic bone
5. Vomer
6. Frontal bone
7. Parietal bone
8. Nasal bone
9. Inferior concha
10. Maxilla
11. Mandible

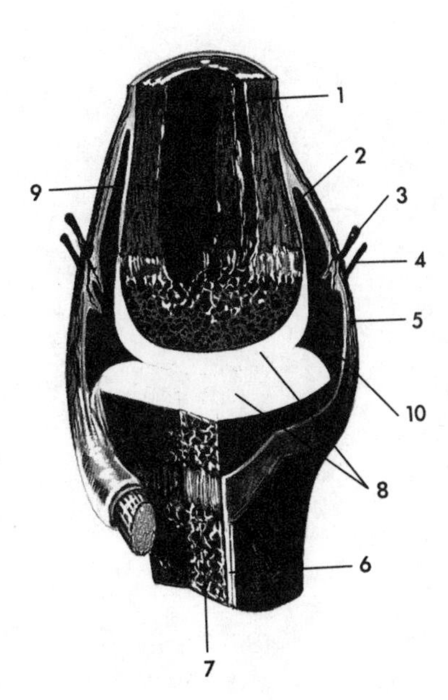

Structure of a Diarthrotic Joint

1. Bone
2. Synovial membrane
3. Blood vessel
4. Nerve
5. Joint capsule
6. Periosteum
7. Bone
8. Articular cartilage
9. Bursa

CHAPTER 6
THE MUSCULAR SYSTEM

Select the correct term

1. A, p. 108
2. B, p. 108
3. C, p. 108
4. C, p. 108
5. A, p. 108
6. B, p. 108
7. C, (may also be B), p. 108
8. A, p. 108
9. C, p. 108
10. C, p. 108

Matching

Group A

11. D, p. 108
12. B, p. 108
13. A, p. 108
14. E, p. 108
15. C, p. 108

Group B

16. E, p. 109
17. C, p. 109
18. B, p. 111
19. A, p. 109
20. D, p. 109

Fill in the blanks

21. Pulling, p. 111
22. Insertion, p. 111
23. Insertion, origin, p. 111
24. Prime mover, p. 111
25. Antagonist, p. 111
26. Synergist, p. 111
27. Tonic contraction, p. 112
28. Muscle tone, p. 112
29. Hypothermia, p. 112
30. ATP, p. 112

True or false

31. Neuromuscular junction, p. 113
32. T
33. T
34. Oxygen debt, p. 112
35. "All or none", p. 114
36. Lactic acid, p. 112
37. Muscle tone, p. 114
38. T
39. Skeletal muscles, p. 112
40. T

Multiple choice

41. A, p. 115
42. B, p. 115
43. B, p. 116
44. C, p. 116
45. D, p. 116
46. A, p. 114
47. B, p. 114
48. C, p. 115
49. B, p. 116
50. D, p. 116

Select the best choice

51. C, p. 117
52. F, p. 117; A and D, p. 124
53. F, p. 117 and B, p. 124
54. A, p. 117
55. C, p. 124
56. B, p. 124 and F, p. 117
57. A, p. 117
58. A and D, p. 124
59. B, p. 124
60. B, p. 124
61. A, p. 120
62. B, p. 117
63. D, p. 117

Multiple choice

64. A, p. 122
65. D, p. 123
66. C, p. 123
67. A, p. 123
68. D, p. 123
69. C, p. 125

APPLYING WHAT YOU KNOW

70. Bursitis
71. Deltoid area
72. Tendon

73. WORD FIND

```
. . . . . . N I G I R O I R . . . . . .
. . . . . . . . . . . S O N . . . . N .
. . . . M U S C L E O T O . R . . T O .
. S . . . . . . . T C I . . O . . E I .
. P . . . . . . O U X . . D T . . N X .
. E . . . . . N D E . . . E A . . O E .
. C . . . . I B L . . . . T T . S S L .
. I . . . C A F . . . . M A O B U Y F N
. B . . . . I . . . Y E G I R U I N . O
. . . . . S . . S . H X A R . R Z O S I
. . . . R . . . O . P T R T . S E V P T
. . . O . . . . L . O E H S . A P I E R
. . D . . . . . E . R N P . . . A T C E
. . . . . . . . U . T S A . . . R I I S
. . . . . . . . S . A I I . . . T S R N
. . . . . . . . . . . O D . . . . . T I
. . T S I G R E N Y S N D I O T L E D .
I S O M E T R I C . . . E U G I T A F .
. . . . S U I M E N C O R T S A G . . .
. . S G N I R T S M A H . T E N D O N .
```

CROSSWORD

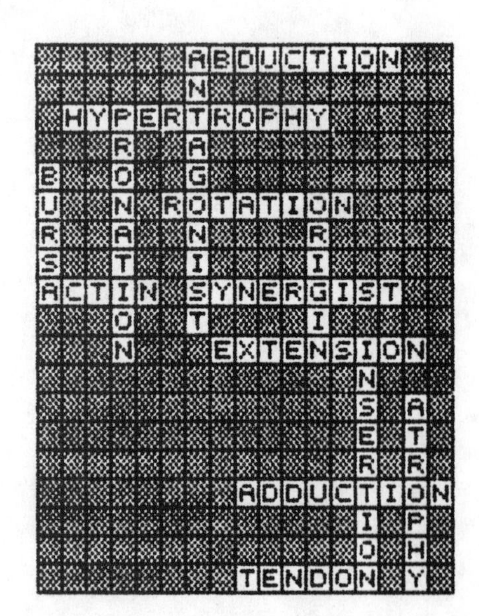

Muscles - Anterior View

1. Sternocleidomastoid
2. Trapezius
3. Pectoralis major
4. Rectus abdominis
5. External abdominal oblique
6. Iliopsoas
7. Quadriceps group
8. Tibialis anterior
9. Peroneus longus
10. Peroneus brevis
11. Soleus
12. Gastrocnemius
13. Sartorius
14. Adductor group
15. Brachialis
16. Biceps brachii
17. Deltoid
18. Facial muscles

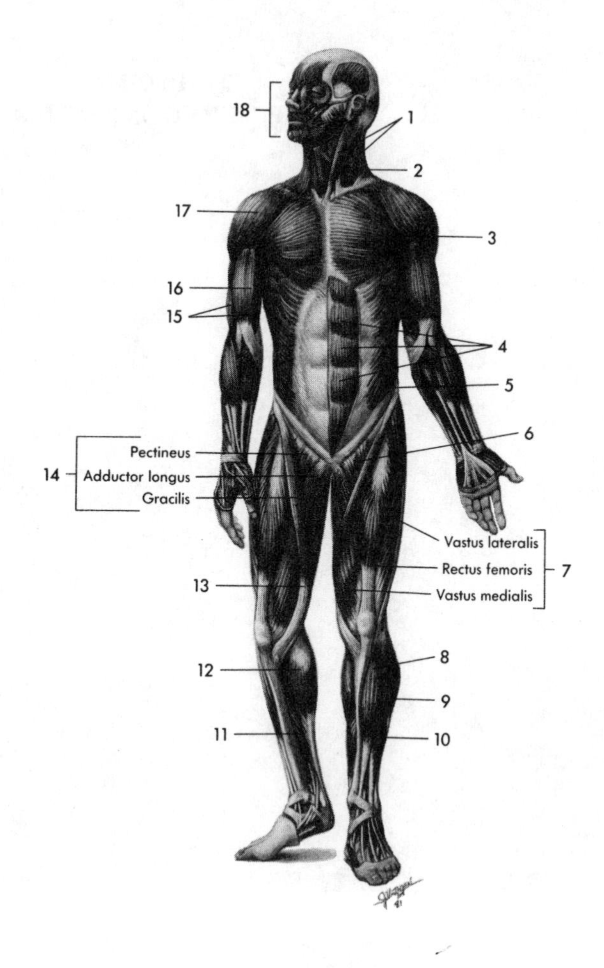

Muscles - Posterior View

1. Trapezius
2. External abdominal oblique
3. Gluteus maximus
4. Adductor magnus
5. Soleus
6. Peroneus brevis
7. Peroneus longus
8. Gastrocnemius
9. Hamstring group
10. Latissimus dorsi
11. Triceps brachii
12. Deltoid
13. Sternocleidomastoid

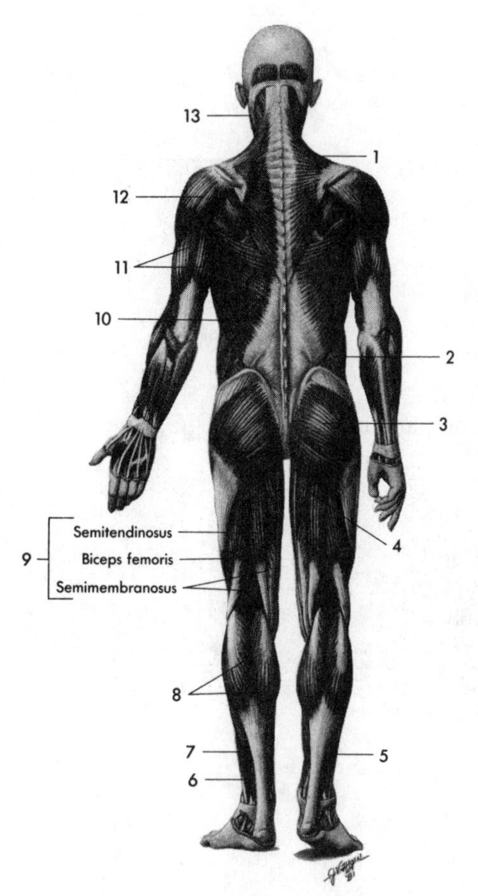

CHAPTER 7
THE NERVOUS SYSTEM

MATCHING

Group A

1. B, p. 132
2. C, p. 132
3. D, p. 132
4. A, p. 132

Group B

5. B, p. 132
6. D, p. 133
7. C, p. 132
8. A, p. 133
9. F, p. 136
10. E, p. 136

Select the best choice

11. A, p. 132
12. B, p. 133
13. B, p. 135
14. A, p. 132
15. A, p. 132
16. B, p. 135
17. B, p. 135
18. A, p. 132
19. B, p. 135
20. A, p. 133

Fill in the blanks

21. Two-neuron arc, p. 137
22. Sensory, interneurons, and motor neurons, p. 137
23. Receptors, p. 137
24. Synapse, p. 138
25. Reflex, p. 138
26. Withdrawal reflex, p. 138
27. Saltatory conduction, p. 138
28. Interneurons, p. 138
29. "Knee jerk", p. 138
30. Gray matter, p. 138

Circle the correct word

31. Do not, p. 138
32. Increases, p. 138
33. Excess, p. 138
34. Postsynaptic, p. 140
35. Presynaptic, p. 140
36. Neurotransmitter, p. 141
37. Communicate, p. 140
38. Specifically, p. 141
39. Sleep, p. 141
40. Pain, p. 141

Multiple choice

41. E, p. 142
42. D, p. 142
43. A, p. 142
44. E, p. 142
45. E, p. 142
46. D, p. 144
47. B, p. 144
48. E, p. 146
49. B, p. 146
50. D, p. 146
51. D, p. 144
52. B, p. 144
53. A, p. 146
54. D, p. 144
55. C, p. 144

True or false

56. 17 to 18 inches, p. 146
57. Bottom of the first lumbar vertebra, p. 148
58. Lumbar punctures, not CAT scan, p. 154
59. Spinal tracts, not dendrites, p. 148
60. T
61. One general function, not several, p. 148
62. Anesthesia, not paralysis, p. 149

Circle the one that does not belong

63. Ventricles (all others refer to meninges)
64. CSF (all others refer to the arachnoid)
65. Pia mater (all others refer to the cerebrospinal fluid)
66. Choroid plexus (all others refer to the dura mater)
67. Brain tumor (all others refer to a lumbar puncture)

CRANIAL NERVES

68.

NERVE	CONDUCT IMPULSES	FUNCTION
I Olfactory		
II		Vision
III	From brain to eye muscles	
IV Trochlear		
V		Sensations of face, scalp, and teeth, chewing movements
VI	From brain to external eye muscles	
VII		Sense of taste; contraction of muscles of facial expression
VIII Acoustic		
IX	From throat and taste buds of tongue to brain; also from brain to throat muscles and salivary glands	
X Vagus		
XI		Shoulder movements; turning movements of head
XII Hypoglossal		

Select the best choice

69. A, p. 152
70. B, p. 154
71. A, p. 153
72. B, p. 156
73. B, p. 153
74. A, p. 153
75. B, p. 153
76. B, p. 154

Matching

77. D, p. 156
78. E, p. 156
79. F, p. 156
80. B, p. 156
81. A, p. 156
82. C, p. 156

Multiple choice

83. C, p. 159
84. B, p. 159
85. B, p. 159
86. D, p. 159
87. A, p. 159
88. A, p. 159

Choose the correct response

89. B, p. 160
90. A, p. 160
91. A, p. 160
92. B, p. 160
93. A, p. 160
94. B, p. 160
95. A, p. 160
96. A, p. 160
97. B, p. 160
98. B, p. 160

Fill in the blanks

99. Acetylcholine, p. 160
100. Adrenergic fibers, p. 160
101. Cholinergic fibers, p. 160
102. Homeostasis, p. 160
103. Heart rate, p. 161
104. Decreased, p. 161

Unscramble the words

105. Neurons
106. Synapse
107. Autonomic
108. Smooth muscle
109. Sympathetic

APPLYING WHAT YOU KNOW

110. Right
111. Hydrocephalus
112. Sympathetic
113. Parasympathetic
114. Sympathetic; No, the digestive process is not active during sympathetic control. Bill may experience nausea, vomiting or discomfort because of this factor. See p. 160.

CROSSWORD

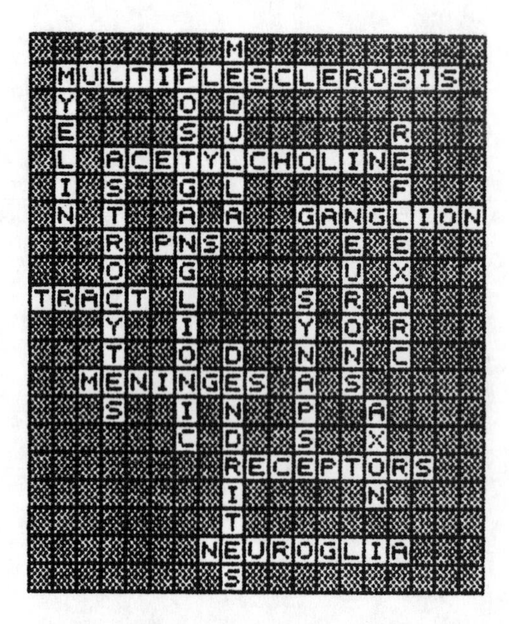

CROSSWORD

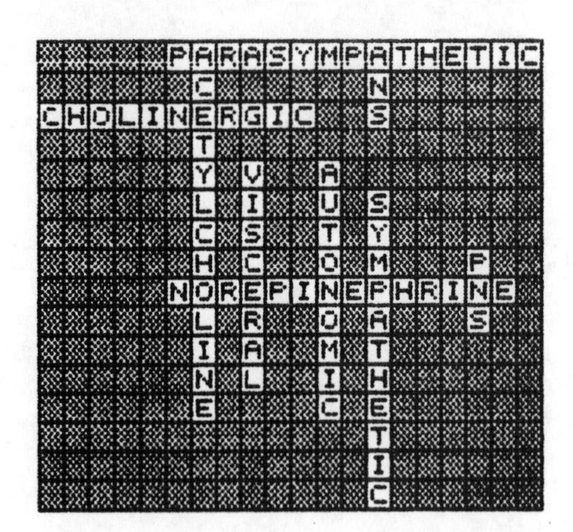

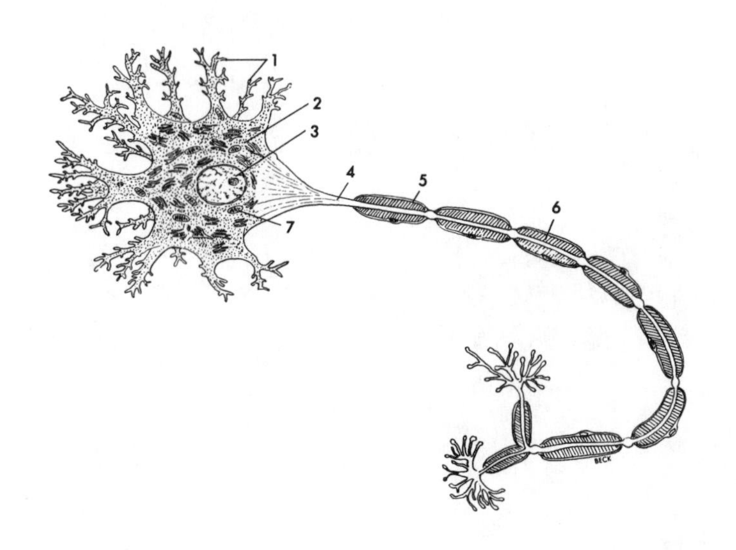

Neuron

1. Dendrites
2. Cell body
3. Nucleus
4. Axon
5. Schwann cell
6. Myelin
7. Mitochondrion

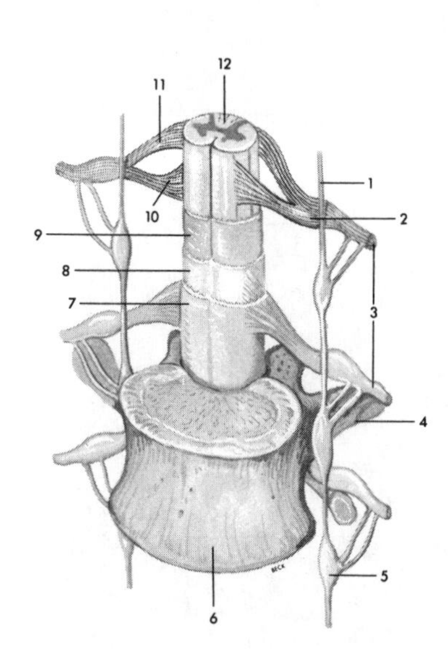

Spinal Cord

1. Sympathetic trunk
2. Spinal ganglion
3. Spinal nerves
4. Transverse process
5. Sympathetic ganglion
6. Body of vertebra
7. Dura mater
8. Arachnoid
9. Pia mater
10. Anterior root
11. Posterior root
12. Spinal cord

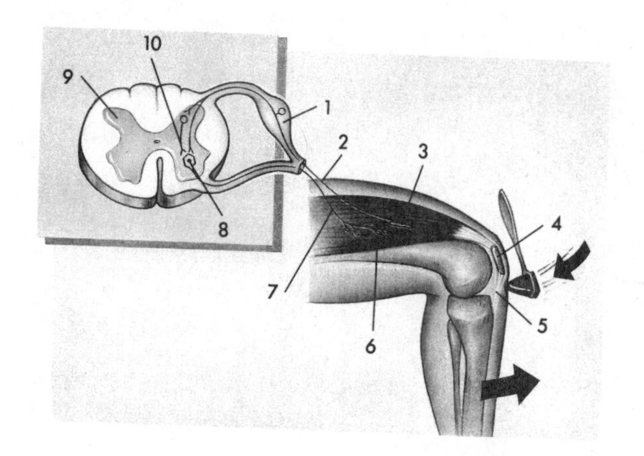

Patellar Reflex

1. Dorsal root ganglion
2. Sensory neuron
3. Stretch receptor
4. Patella
5. Patellar tendon
6. Quadriceps muscle
7. Motor neuron
8. Monosynaptic synapse
9. Gray matter
10. Interneuron

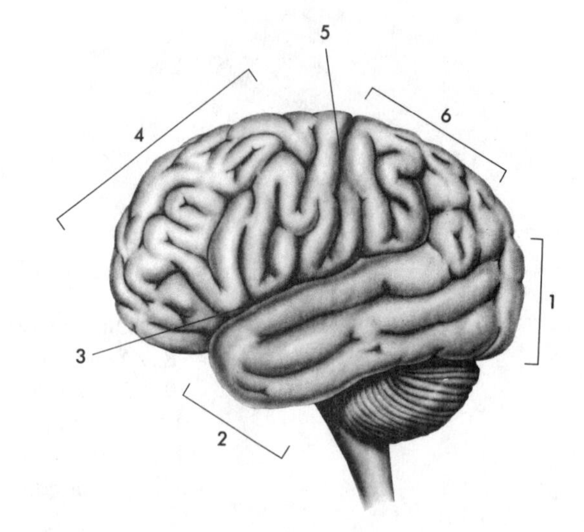

The Cerebrum

1. Occipital lobe
2. Temporal lobe
3. Fissure of Sylvius
4. Frontal lobe
5. Fissure of Rolando
6. Parietal lobe

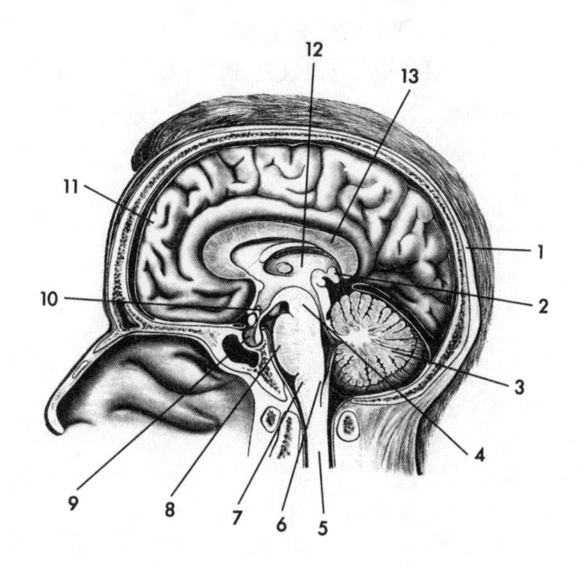

Sagittal Section of the Central Nervous System

1. Skull
2. Pineal gland
3. Cerebellum
4. Midbrain
5. Spinal cord
6. Medulla
7. Reticular formation
8. Pons
9. Pituitary gland
10. Hypothalamus
11. Cerebral cortex
12. Thalamus
13. Corpus callosum

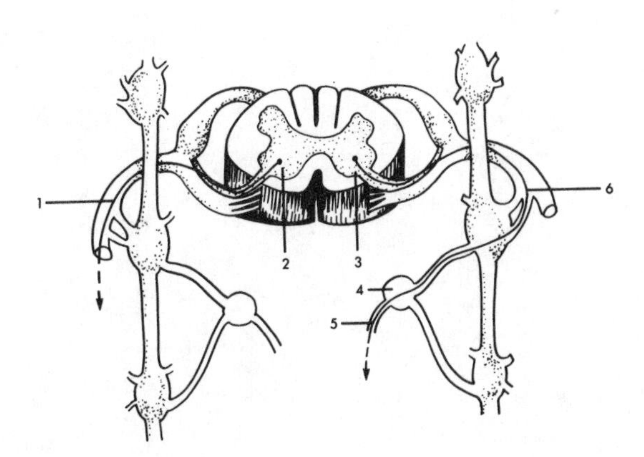

Neuron Pathways

1. Somatic motor neuron's axon
2. Cell body of somatic motor neuron
3. Cell body of preganglionic neuron
4. Collateral ganglion
5. Postganglionic neuron's axon
6. Preganglionic sympathetic neuron's axon

CHAPTER 8
SENSE ORGANS

Matching

1. D, p. 169
2. B, p. 178
3. A, p. 169
4. E, p. 178
5. C, p. 169

Multiple choice

6. C, p. 168
7. E, p. 168
8. B, p. 168
9. C, p. 168
10. E, p. 168
11. D, p. 168
12. A, p. 170
13. B, p. 170
14. C, p. 170
15. B, p. 170
16. D, p. 170
17. A, p. 172
18. D, p. 171

Select the best choice

19. B, p. 174
20. C, p. 174
21. B, p. 174
22. A, p. 174
23. C, p. 174
24. A, p. 173
25. C, p. 174
26. B, p. 174
27. A, p. 174
28. C, p. 175

Fill in the blanks

29. Auricle and external auditory canal, p. 173
30. Eardrum, p. 174
31. Ossicles, p. 174
32. Oval window, p. 174
33. Otitis media, p. 174
34. Vestibule, p. 174
35. Mechanoreceptors, p. 174
36. Crista ampullaris, p. 174

Circle the correct answer

37. Papillae, p. 176
38. Cranial, p. 176
39. Mucus, p. 176
40. Memory, p. 178
41. Proprioception, p. 178

APPLYING WHAT YOU KNOW

42. External otitis
43. Cataracts
44. The eustachian tube connects the throat to the middle ear and provides a perfect pathway for the spread of infection
45. Hydrocephalus

46. **WORD FIND**

```
. . . . . . . A V I T C N U J N O C . . .
. . . . . . S U C N I . . . . . . . . . .
O L F A C T O R Y . . . . . . . N . . . .
. . . . . . . . . . . . . . P A . . . . .
. C A T A R A C T S . . . H . I . . . . .
G U S T A T O R Y . . . O . . H . . . . .
. . . P A P I L L A E T A R . C . . . . .
. . . . . . . . . . O I E . . A . . . . .
. . C O C H L E A P P C . S . T . . . . .
. C E R U M E N I O E . . E . S . . . . .
. . . . . . . G Y P . . . N . U . . . . .
. C O N E S M B T . . . . S . E . . . . .
. . . . . E S O A I P O R E P Y H . . . .
. . . . N E R . . . . . . S . . . . . . .
. . . T R S . . . . . . . . . . . . . . .
. . . P M E C H A N O R E C E P T O R S
N O I T C A R F E R . . . . . . E Y E . .
. . . . . . . . . . . . . . . . . . . . .
. . . . . . . R O D S . . . . . . . . . .
. . . . . . . . . . . . . . . . . . . . .
```

CROSSWORD

						V												
T	Y	M	P	A	N	I	C	M	E	M	B	R	A	N	E			
						T												
						R												
C	R	I	S	T	A	E	A	M	P	U	L	L	A	R	I	S		
	E					O												
	T					U												
	I					S	C	L	E	R	A					C		
	N					H		E			Q			P		O		
	A					U		N			U			U		C		
						M		S			E			P		H		
						O					O	S	S	I	C	L	E	S
			C	H	O	R	O	I	D		U			L		E		
											S					A		
			E	N	D	O	L	Y	M	P	H							
											U							
											M							
											O							
									A	U	R	I	C	L	E			

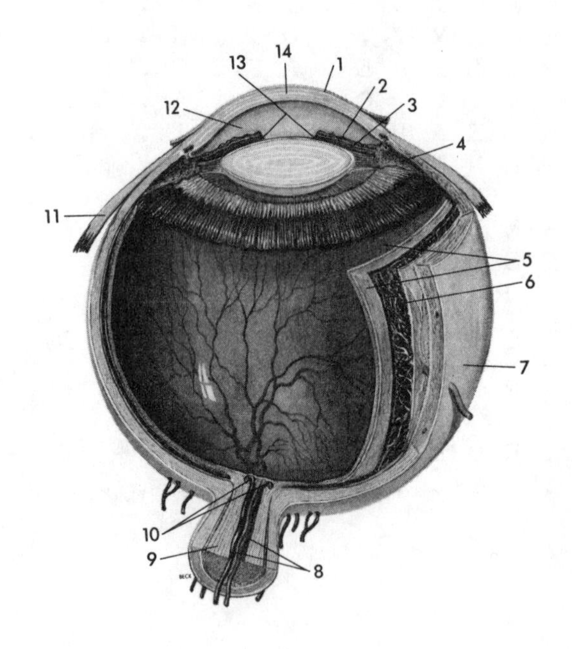

Eye

1. Conjunctiva
2. Iris
3. Posterior cavity
4. Ciliary muscle
5. Retina
6. Choroid layer
7. Sclera
8. Central retinal artery and vein
9. Optic nerve
10. Optic disc (blind spot)
11. Medial rectus muscle
12. Anterior cavity
13. Pupil
14. Cornea

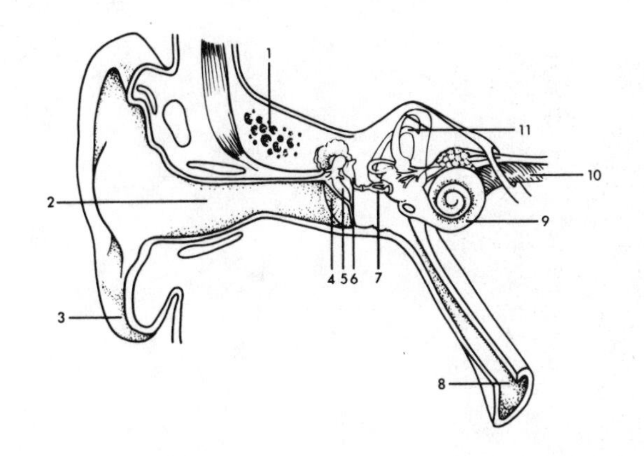

Ear

1. Temporal bone
2. External auditory canal
3. Auricle (Pinna)
4. Tympanic membrane (eardrum)
5. Malleus
6. Incus
7. Stapes
8. Auditory (Eustachian) tube
9. Cochlea
10. Cochlear nerve
11. Semicircular canals

CHAPTER 9
ENDOCRINE SYSTEM

Matching

Group A

1. D, p. 185
2. C, p. 185
3. E, p. 185
4. A, p. 185
5. B, p. 185

Group B

6. E, p. 188
7. C, p. 189
8. A, p. 184
9. D, p. 184
10. B, p. 184

Fill in the blanks

11. Second messenger, p. 184
12. Recognize, p. 184
13. First messengers, p. 184

14. Target organs, p. 184
15. Cyclic AMP, 184
16. Target cells, p. 186
17. Steroid abuse, p. 186

Multiple choice

18. B, p. 190
19. E, p. 190
20. D, p. 190
21. D, p. 190
22. A, p. 190
23. C, p. 190
24. C, p. 192
25. B, p. 190
26. A, p. 190
27. A, p. 190
28. D, p. 192
29. B, p. 192
30. A, p. 193
31. C, p. 193
32. C, p. 193

Select the best answer

33. A, p. 190
34. B, p. 190
35. B, p. 193
36. C, p. 193
37. A, p. 191
38. C, p. 193
39. A, p. 190
40. A, p. 190
41. A, p. 191
42. C, p. 193

Circle the correct term

43. Below, p. 193
44. Calcitonin, p. 193
45. Iodine, p. 193
46. Do not, p. 193
47. Thyroid, p. 193
48. Decrease, p. 193
49. Hypothyroidism, p. 195
50. Cretinism, p. 195
51. PTH, p. 195
52. Increase, p. 195

Fill in the blanks

53. Adrenal cortex and adrenal medulla, p. 196
54. Corticoids, p. 197
55. Mineralocorticoids, p. 197
56. Glucocorticoids, p. 197
57. Sex hormones, p. 197
58. Gluconeogenesis, p. 198
59. Blood pressure, p. 198
60. Epinephrine and norepinephrine, p. 198
61. Stress, p. 199
62. General adaptation syndrome, p. 199

Select the best response

63. A, p. 200
64. A, p. 198
65. B, p. 199
66. A, p. 200
67. B, p. 199
68. A, p. 198
69. A, p. 198

Circle the term that does not belong

70. Beta cells (all others refer to glucagon)
71. Glucagon (all others refer to insulin)
72. Thymosin (all others refer to female sex glands)
73. Chorion (all others refer to male sex glands)
74. Aldosterone (all others refer to the thymus gland)
75. ACTH (all others refer to the placenta)
76. Semen (all others refer to the pineal gland)

APPLYING WHAT YOU KNOW

77. She was pregnant
78. Zona reticularis of the adrenal cortex

79. WORD FIND

```
. . . . H Y P E R C A L C E M I A . . .
. . . . . . . A R . . . . E . . . . . .
. . . . . . M . E . . D . N . . . . . .
. . . . . E . . T . . I . O . . . . . .
. . . . D . . . I . . A . M . . . . . .
. . . E . N . . O . . B . R . . . . . .
. . X . . . O . G . . E . O . . . . . .
. Y . . . . . G . . . T . H . . E . . .
M . . . . . . . A . . E . . . D N . . H
. . . . . . . . . C . S . . I . D . . Y
C R E T I N I S M . U . . U . . O . . P
E X O C R I N E . . . L R . . . C . . O
. . . . . . . . . . . E G . . . R . . G
. . . . . . . . . . S . . . . . I . . L
. . . . . . . . . I . . . . . . N . . Y
. . . . . . . . S . . . . . . . E . . C
. . S D I O C I T R O C . . . . . . . E
. S T R E S S S D I O R E T S . . . . M
. . . . N O I T Z A I N I E T U L . . I
. P R O S T A G L A N D I N S . . . . A
```

CROSSWORD

```
###M################
###S################
CUSHINGSSYNDROME####
###########I########
HYPOGLYCEMIA##GOITER
##T#L######B###X####
ADH#U#ACROMEGALY####
####C#C####T###T####
####O#T####E###O####
####G#H###FSH##C####
####E######M###I####
#EPINEPHRINE##ANH###
####E######L########
####S######L########
####I######I########
####S######T########
###########U########
###########S########
```

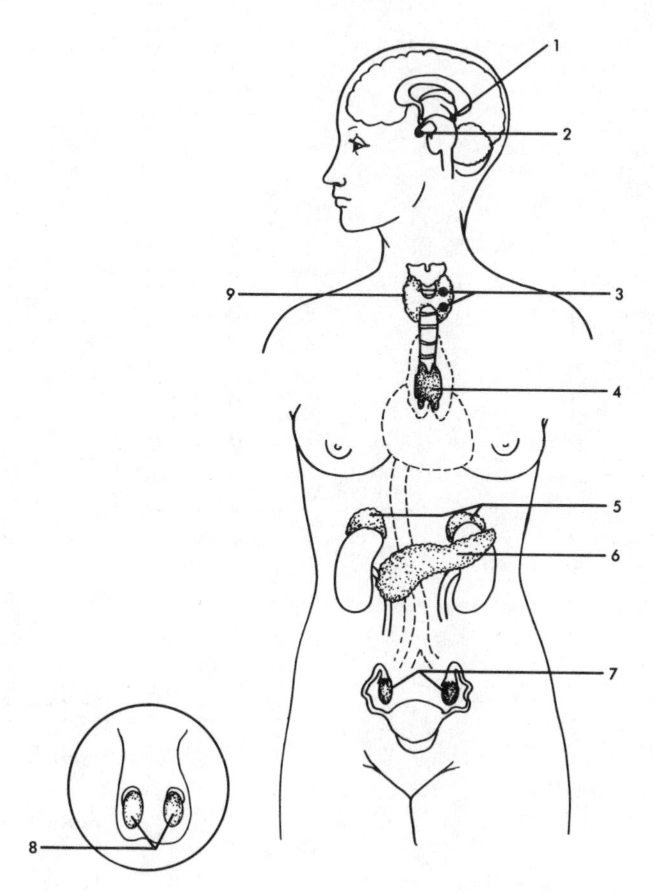

Endocrine Glands

1. Pineal
2. Pituitary
3. Parathyroids
4. Thymus
5. Adrenal
6. Pancreas
7. Ovaries
8. Testes
9. Thyroid

CHAPTER 10
BLOOD

Multiple choice

1. B, p. 210
2. C, p. 210
3. A, p. 211
4. B, p. 212
5. D, p. 210
6. C, p. 214
7. D, p. 212
8. A, p. 216
9. B, p. 218
10. B, p. 215
11. B, p. 210
12. E, p. 210
13. B, p. 210
14. C, p. 210

15. D, p. 210
16. E, p. 212
17. D, p. 212
18. B, p. 214
19. D, p. 211
20. B, p. 214
21. C, p. 212 and 214
22. B, p. 212
23. A, p. 215
24. E, p. 215
25. E, p. 216
26. D, p. 216
27. D, p. 216

Fill in the blank areas

28. p. 217

Blood Type	Antigen	Antibody
A	A	Anti-B
B	B	Anti-A
AB	A, B	None
O	None	Anti-A, Anti-B

Fill in the blanks

29. Antigen, p. 217
30. Antibody, p. 217
31. Agglutinate, p. 217
32. Erythroblastosis fetalis, p. 218
33. Rhesus monkeys, p. 218

APPLYING WHAT YOU KNOW

34. No. If Mrs. Payne were a negative Rh factor and her husband were a positive Rh factor, it would set up the strong possibility of erythroblastosis fetalis.

35. Both procedures assist the clotting process.

36. **WORD FIND**

```
. . . S U B M O R H T . . . . . . . . .
. . R . B H E . P H A G O C Y T E S L .
. . E . A E M . . . . . . . . . . . E .
L . C . S M B . . . T H R O M B I N U .
E . I . O O O . . . . . . . S . M . K .
U . P . P G L . . . . . . E . O . . E .
K . I . H L U . A . . . T F N . . . M .
O . E . I O S . I . . Y A O . . . . I .
C S N . L B . . D . C C C . . . . . A .
Y U T . . I . . S O T Y . . . . . . . .
T S . R O N O D R O T . A . . . . . . .
E E . . . . . H R E . N . . . . . . . .
S H N . . . T . . . T A . . . . . . . .
. R I . . Y . . . I . N A M S A L P . .
. . R . R . . . G . . T . E P Y T . . S
. . B E . . . E . . . I . . . . . . E .
. . I . . . N . . . . B . . . . . R . .
. . F . . A I M E N A O . . . . U . . .
N I R A P E H . . . . D . . . M . . . .
H E M A T O C R I T . Y . . . . . . . .
```

CROSSWORD

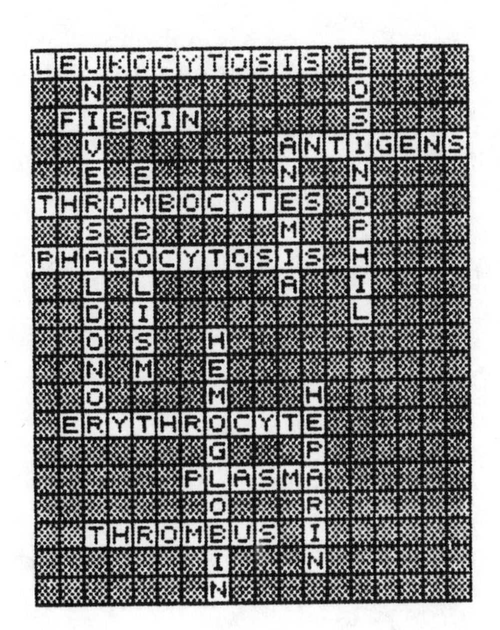

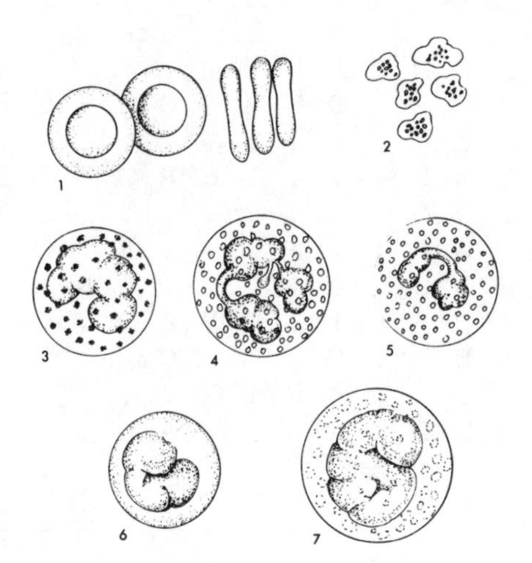

Human Blood Cells

1. Red blood cells
2. Platelets
3. Basophil
4. Neutrophil
5. Eosinophil
6. Lymphocyte
7. Monocyte

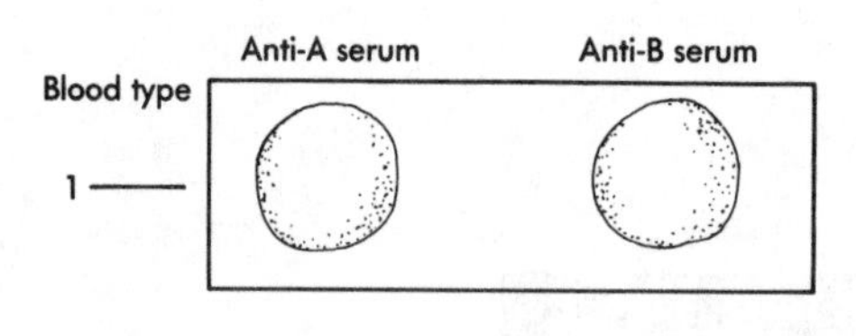

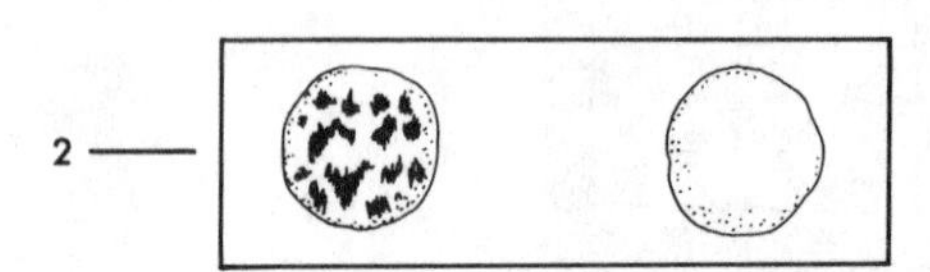

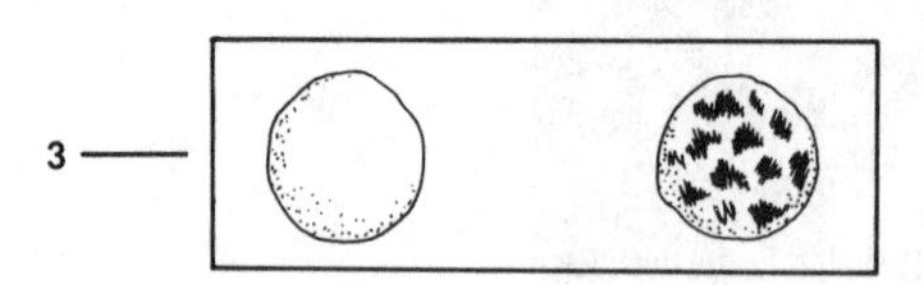

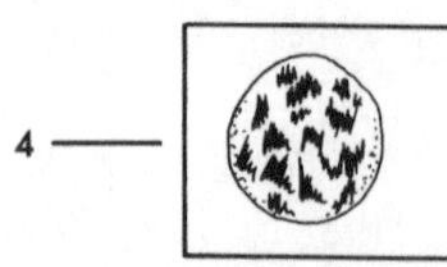

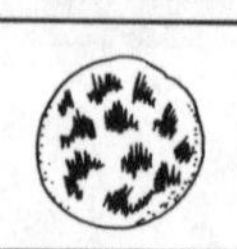

Normal blood

Agglutinated blood

Blood Types

1. O
2. A
3. B
4. AB

CHAPTER 11
THE CIRCULATORY SYSTEM

Fill in the blanks

1. CPR, p. 224
2. Interatrial septum, p. 224
3. Atria, p. 224
4. Ventricles, p. 224
5. Myocardium, p. 224
6. Endocarditis, p. 227
7. Bicuspid or mitral and tricuspid, p. 227
8. Pulmonary circulation, p. 229
9. Coronary embolism or coronary thrombosis, p. 229
10. Myocardial infarction, p. 229
11. Sinoatrial node, p. 231
12. P, QRS complex and T, p. 233
13. Repolarization, p. 233

Select the best answer

14. A, p. 227
15. K, p. 224
16. G, p. 229
17. C, p. 227
18. D, p. 227
19. F, p. 224
20. H, p. 229
21. B, p. 229
22. E, p. 233
23. I, p. 227
24. J, p. 233
25. L, p. 224
26. M, p. 227

Matching

27. D, p. 233
28. B, p. 233
29. C, p. 233
30. G, p. 233
31. A, p. 235
32. E, p. 233
33. F, p. 233

Multiple choice

34. D, p. 233
35. C, p. 233
36. B, p. 235
37. A, p. 233
38. B, p. 235
39. B, p. 239
40. D, p. 241
41. B, p. 243
42. B, p. 243
43. A, p. 243
44. D, p. 233
45. A, p. 235

True or false

46. Highest in arteries, lowest in veins, p. 243
47. Blood pressure gradient, p. 243
48. Stop, p. 243
49. High, p. 244
50. Decreases, p. 244
51. T
52. T
53. T
54. Stronger will increase, weaker will decrease, p. 245
55. Contract, p. 245
56. Relax, p. 245
57. Artery, p. 246
58. T
59. T
60. Brachial, p. 246

Unscramble the words

61. Systemic
62. Venule
63. Artery
64. Pulse
65. Vessel

APPLYING WHAT YOU KNOW

66. Coronary bypass surgery
67. Artificial pacemaker
68. The endocardial lining can become rough and abrasive to red blood cells passing over its surface. As a result, a fatal blood clot may be formed.

CROSSWORD

	V																
	E		A							H							
M	I	T	R	A	L	V	A	L	V	E		E					
	N		T							P		N					
			E			M	Y	O	C	A	R	D	I	U	M		
	C	P	R							T		O					
			I							I		C					
	E		O		V					C		A	T	R	I	U	M
	P	U	L	S	E					P		R					
	I		E		N					O		D					
	C				T					R		I					
	A	R	T	E	R	Y				T		T			P		
	R				I				C	A	P	I	L	L	A	R	Y
	D				C					L		S			C		
	I				L										E		
	U				E										M		
	M				S										A		
															K		
															E		
															R		

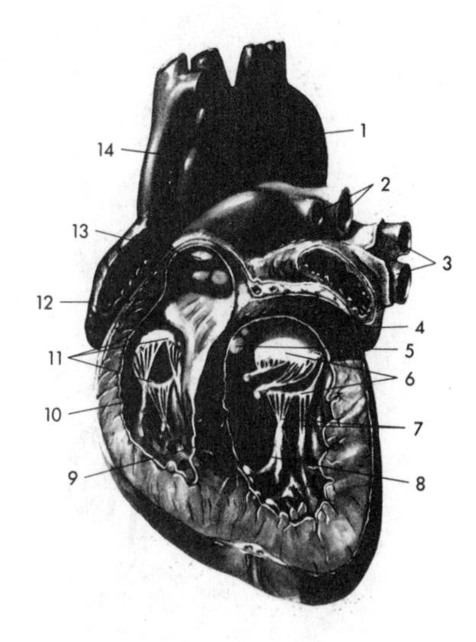

The Heart

1. Aorta
2. Pulmonary arteries
3. Left pulmonary veins
4. Left atrium
5. Aortic semilunar valve
6. Bicuspid valve
7. Chordae tendineae
8. Left ventricle
9. Interventricular septum
10. Right ventricle
11. Tricuspid valve
12. Right atrium
13. Pulmonary semilunar valve
14. Superior vena cava

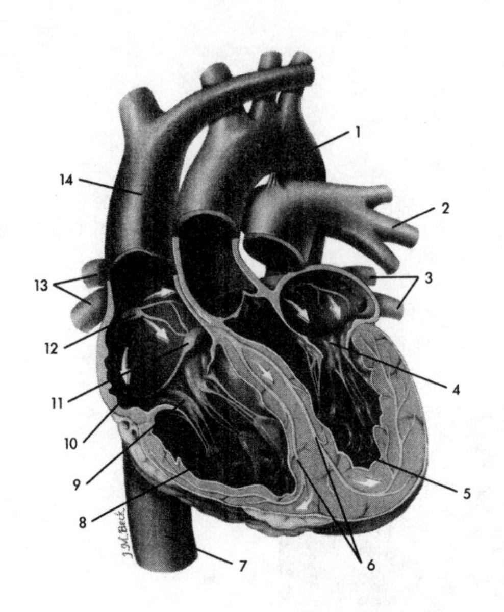

Conduction System of the Heart

1. Aorta
2. Pulmonary artery
3. Pulmonary veins
4. Mitral (bicuspid) valve
5. Left ventricle
6. Right and left branches of AV bundle
7. Inferior vena cava
8. Right ventricle
9. Tricuspid valve
10. Right atrium
11. Atrioventricular node (AV node)
12. Sinoatrial node (SA node or pacemaker)
13. Pulmonary veins
14. Superior vena cava

Fetal Circulation

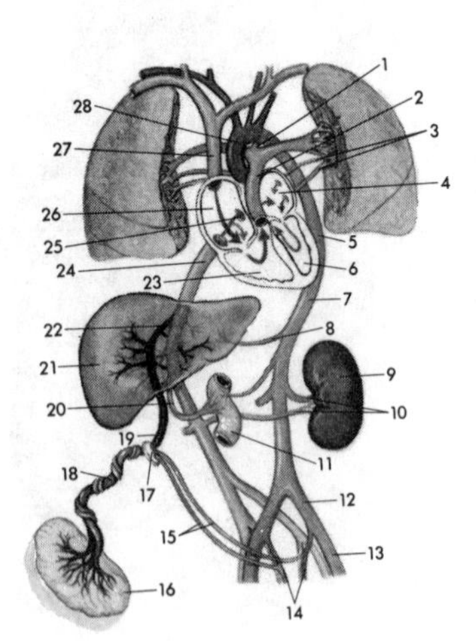

1. Ductus arteriosus
2. Pulmonary artery
3. Pulmonary veins
4. Left atrium
5. Thoracic aorta
6. Left ventricle
7. Abdominal aorta
8. Hepatic artery
9. Kidney
10. Renal vein and artery
11. Intestine
12. Left common iliac artery
13. External iliac artery
14. Internal iliac arteries
15. Umbilical arteries
16. Placenta
17. Fetal umbilicus
18. Umbilical cord
19. Umbilical vein
20. Portal vein
21. Liver
22. Ductus venosus
23. Right ventricle

24. Inferior vena cava
25. Foramen ovale
26. Right atrium
27. Superior vena cava
28. Ascending aorta

Hepatic Portal Circulation

1. Inferior vena cava
2. Stomach
3. Gastric vein
4. Left gastroepiploic vein
5. Spleen
6. Splenic vein with pancreatic branches
7. Tail of pancreas
8. Right gastroepiploic vein
9. Descending colon
10. Inferior mesenteric vein
11. Small intestine
12. Appendix
13. Superior mesenteric vein
14. Ascending colon
15. Head of pancreas
16. Duodenum
17. Hepatic portal vein
18. Cystic vein
19. Liver
20. Hepatic veins

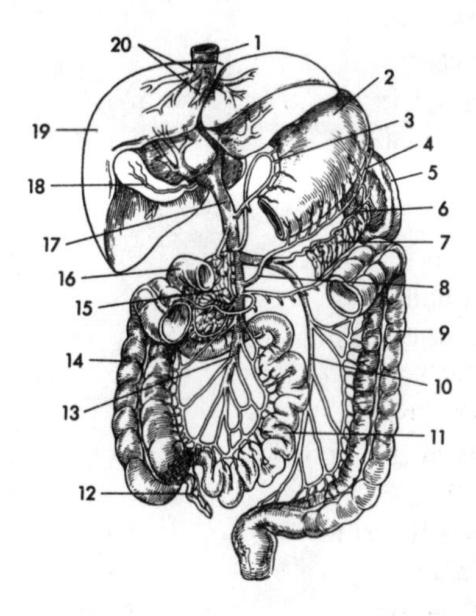

Principal Arteries of the Body

1. Occipital
2. Internal carotid
3. External carotid
4. Left common carotid
5. Left subclavian
6. Arch of aorta
7. Pulmonary
8. Left coronary
9. Aorta
10. Celiac
11. Splenic
12. Renal
13. Inferior mesenteric
14. Radial
15. Ulnar
16. Anterior tibial
17. Popliteal
18. Femoral
19. Deep femoral
20. External iliac
21. Internal iliac
22. Common iliac
23. Abdominal aorta
24. Superior mesenteric
25. Brachial
26. Axillary
27. Right coronary
28. Brachiocephalic
29. Right common carotid
30. Facial

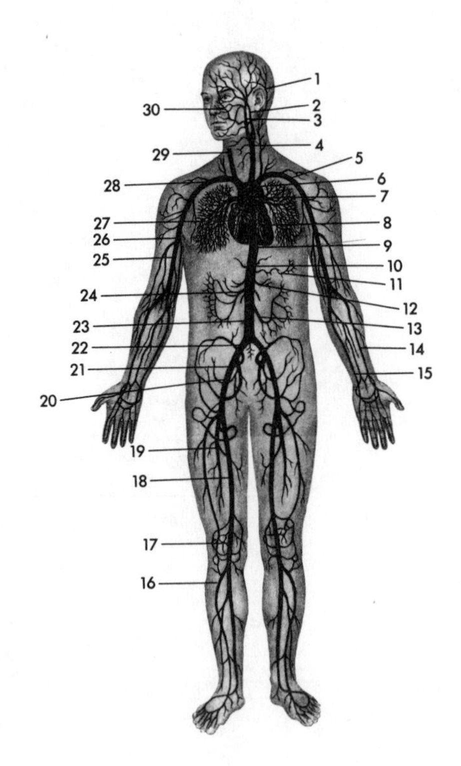

Principal Veins of the Body

1. Superior sagittal sinus
2. External jugular
3. Internal jugular
4. Left brachiocephalic
5. Left subclavian
6. Cephalic
7. Axillary
8. Left coronary
9. Basilic
10. Long thoracic
11. Splenic
12. Inferior mesenteric
13. Common iliac
14. Internal iliac
15. Femoral
16. Popliteal
17. Peroneal
18. Posterior tibial
19. Anterior tibial

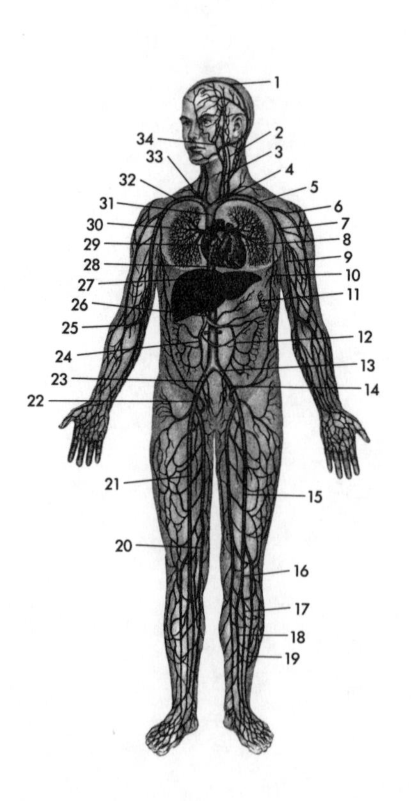

20. Great saphenous
21. Femoral
22. External iliac
23. Common iliac
24. Superior mesenteric
25. Median cubital
26. Portal
27. Hepatic
28. Inferior vena cava
29. Right coronary
30. Pulmonary
31. Superior vena cava
32. Right subclavian
33. Right brachiocephalic
34. Anterior facial

Normal ECG Deflections

1. Atrial depolarization
2. Ventricular depolarization
3. Ventricular repolarization

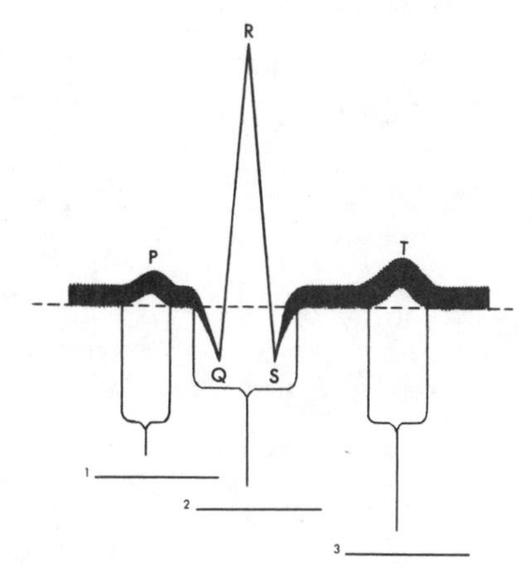

CHAPTER 12
THE LYMPHATIC SYSTEM AND IMMUNITY

Fill in the blanks

1. Lymph, p. 252
2. Interstitial fluid, p. 252
3. Lymphatic capillaries, p. 252
4. Right lymphatic duct and thoracic duct, p. 252
5. Cisterna chyli, p. 252
6. Lymph nodes, p. 254
7. Afferent, p. 254
8. Efferent, p. 254

Choose the correct response

9. B, p. 256
10. C, p. 256
11. C, p. 256
12. A, p. 256
13. C, p. 256
14. A, p. 256
15. A, p. 256

Matching

16. C, p. 257
17. A, p. 257
18. E, p. 257
19. B, p. 257
20. D, p. 257

Choose the correct term

21. C, p. 255
22. D, p. 255
23. E, p. 259
24. A, p. 258
25. H, p. 259
26. B, p. 259
27. I, p. 259
28. F, p. 260
29. J, p. 259
30. G, p. 260

Circle the term that does not belong

31. Antigen (all others refer to antibodies)
32. Complement (all others refer to antigens)
33. Antigen (all others refer to monoclonal antibodies)
34. Complement (all others refer to allergy)
35. Monoclonal (all others refer to complement)

Multiple choice

36. D, p. 262
37. D, p. 264
38. C, p. 264
39. B, p. 263
40. C, p. 265
41. C, p. 264
42. E, p. 265
43. E, p. 265
44. C, p. 265
45. E, p. 265
46. E, p. 266
47. D, p. 261
48. E, p. 261
49. A, p. 265
50. B, p. 266

Fill in the blanks

51. Stem cell, p. 264
52. Activated B cell, p. 265
53. Plasma cells, p. 265
54. Thymus gland, p. 265
55. Azidothymidine or AZT, p. 261
56. AIDS, p. 261
57. AIDS-related complex or ARC syndrome, p. 261

Unscramble the words

58. Complement
59. Immunity
60. Clones
61. Interferon
62. Memory cells

APPLYING WHAT YOU KNOW

63. Interferon would possibly decrease the severity of the chickenpox virus.
64. AIDS
65. Baby Coyle had no means of producing T cells, thus making him susceptible to several diseases. Isolation was a means of controlling his exposure to these diseases.

CROSSWORD

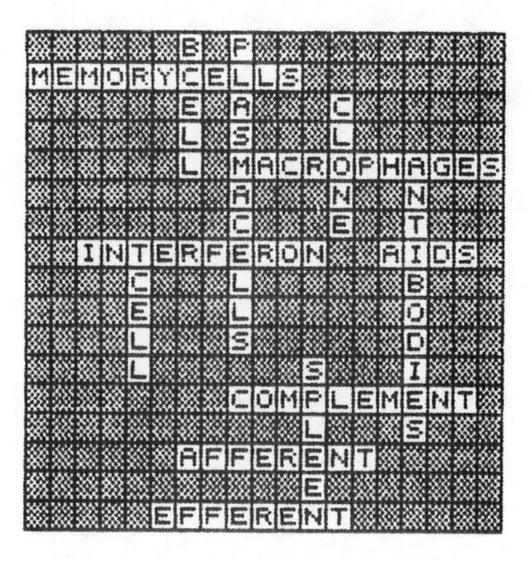

B Cell Development

1. Immature B cells
2. Activated B cells
3. Memory cells
4. Antibodies

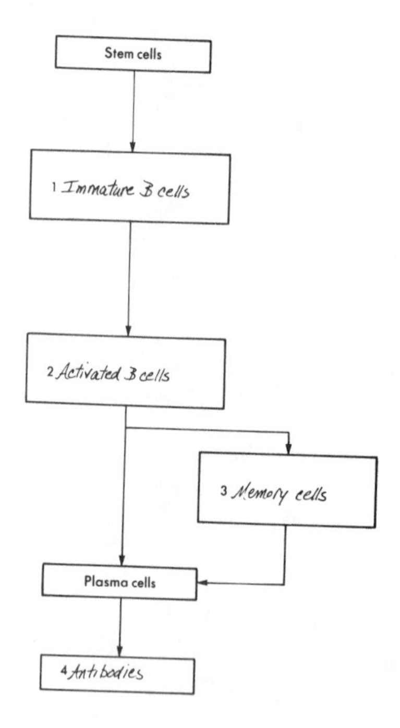

Function of Sensitized T Cells

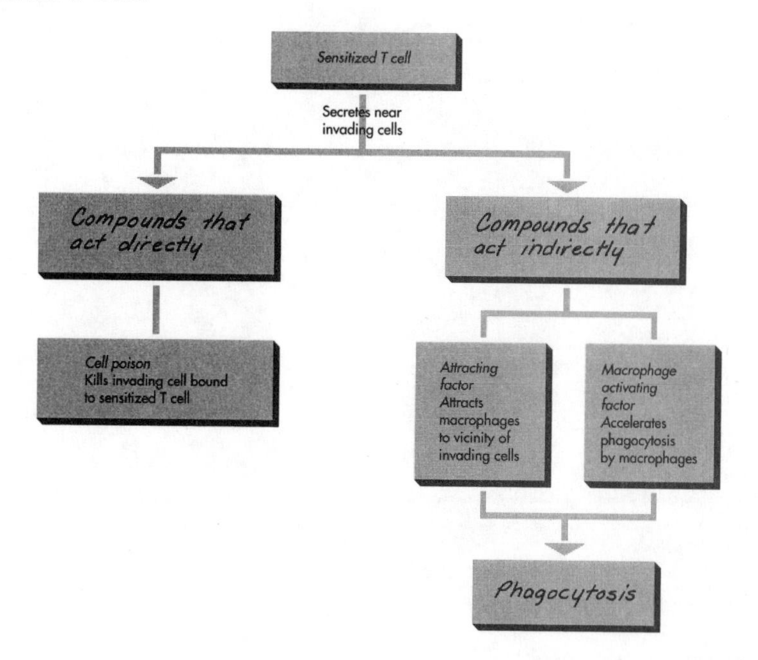

Function of Antibodies

1. Antigen
2. Antibody
3. Antigen
4. Antibody complex
5. Kills invader

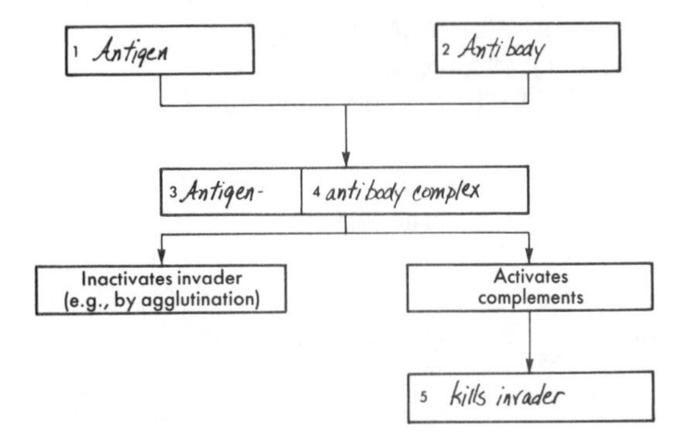

CHAPTER 13
THE RESPIRATORY SYSTEM

Matching

1. J, p. 272
2. G, p. 273
3. A, p. 272
4. I, p. 275
5. B, p. 275
6. F, p. 275
7. C, p. 275
8. H, p. 275
9. D, p. 275
10. E, p. 272

Fill in the blanks

11. Air distributor, p. 272
12. Gas exchanger, p. 272
13. Filters, p. 272
14. Warms, p. 272
15. Humidifies, p. 272
16. Nose, p. 272
17. Pharynx, p. 272
18. Larynx, p. 272
19. Trachea, p. 272
20. Bronchi, p. 272
21. Lungs, p. 272
22. Alveoli, p. 272
23. Diffusion, p. 272
24. Respiratory membrane, p. 272
25. Surface, p. 272

Circle the one that does not belong

26. Oropharynx (the others refer to the nose)
27. Conchae (the others refer to paranasal sinuses)
28. Epiglottis (the others refer to the pharynx)
29. Uvula (the others refer to the adenoids)
30. Larynx (the others refer to the eustachian tubes)
31. Tonsils (the others refer to the larynx)
32. Eustachian tube (the others refer to the tonsils)
33. Pharynx (the others refer to the larynx)

Choose the correct response

34. A, p. 277
35. B, p. 277
36. A, p. 276
37. A, p. 277
38. A, p. 276
39. B, p. 278
40. B, p. 278
41. C, p. 279

Fill in the blanks

42. Trachea, p. 279
43. C-rings of cartilage, p. 279
44. Heimlich maneuver, pp. 280 and 287
45. Primary bronchi, p. 280
46. Alveolar sacs, p. 281
47. Apex, p. 283
48. Pleura, p. 283
49. Pleurisy, p. 283
50. Pneumothorax, p. 283

True or false

51. Breathing, p. 284
52. Expiration, p. 285
53. Down, p. 285 (review Chapter 2)
54. Internal respiration, p. 284
55. T
56. 1 pint, p. 289
57. T
58. Vital capacity, p. 289
59. T

Multiple choice

60. E, p. 284
61. C, p. 285
62. C, p. 285
63. B, p. 284
64. D, p. 289
65. D, p. 289
66. D, p. 289

Matching

67. E, p. 290
68. B, p. 290
69. G, p. 291
70. A, p. 291
71. F, p. 291
72. D, p. 291
73. C, p. 291

Unscramble the words

74. Pleurisy
75. Bronchitis
76. Epistaxis
77. Adenoids
78. Inspiration

APPLYING WHAT YOU KNOW

79. During the day Peter's cilia are paralyzed because of his heavy smoking. They use the time, when Peter is asleep, to sweep accumulations of mucus and bacteria towards the pharynx. When Peter awakes, these collections are waiting to be eliminated.

80. Swelling of the tonsils or adenoids caused by infection may make it difficult or impossible for air to travel from the nose into the throat. The individual may be forced to breathe through the mouth.

CROSSWORD

												P				T			
				T			A	L	V	E	O	L	I			R			
A		E		O								E				A			
P	A	R	A	N	A	S	A	L	S	I	N	U	S			C			
N		V		S								R				H		H	
E				I						S	P	I	R	O	M	E	T	E	R
A				L	A	R	Y	N	X			S				A		I	
				L								Y						M	
			C	E														L	
B	R	O	N	C	H	I												I	
			N	T														C	
			C	O														H	
			H	M															
			A	Y															
			E																

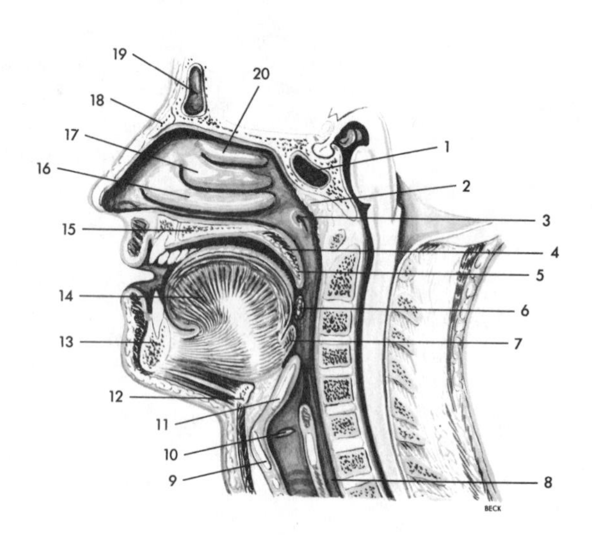

Sagittal View of Face and Neck

1. Sphenoid air sinus
2. Pharyngeal tonsil (adenoids)
3. Auditory tube
4. Soft palate
5. Uvula
6. Palatine tonsil
7. Lingual tonsil
8. Esophagus
9. Thyroid cartilage
10. Vocal cords
11. Epiglottis
12. Hyoid bone
13. Mandible
14. Tongue
15. Hard palate
16. Inferior concha
17. Middle concha
18. Nasal bone
19. Frontal air sinus
20. Superior concha

Respiratory Organs

1. Pharynx
2. Left main bronchus
3. Bronchioles
4. Right main bronchus
5. Trachea
6. Alveolar sacs
7. Alveolar duct
8. Alveolus
9. Capillary

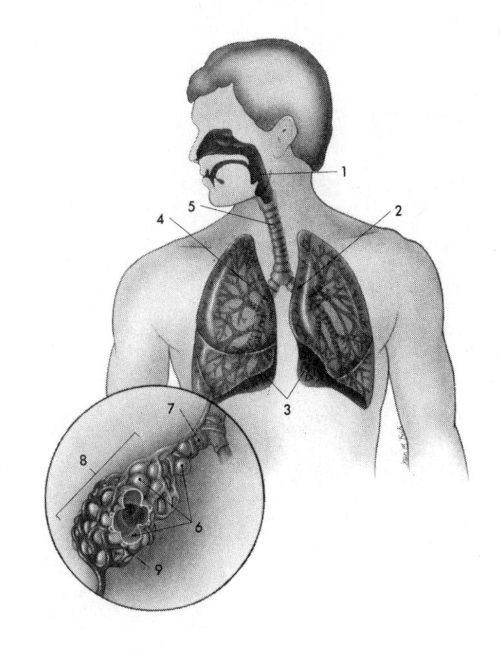

Pulmonary Ventilation Volumes

1. Total lung capacity
2. Inspiratory reserve volume
3. Tidal volume
4. Expiratory reserve volume
5. Residual volume

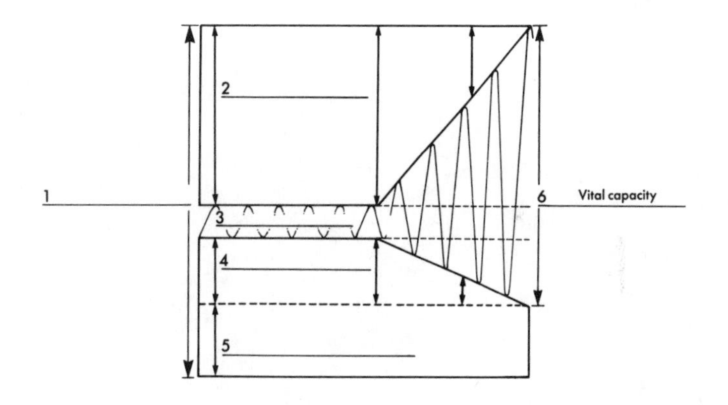

CHAPTER 14
THE DIGESTIVE SYSTEM

Fill in the blanks

1. Gastrointestinal tract or G.I. tract, p. 296
2. Mechanical, p. 296
3. Chemical, p. 296
4. Feces, p. 296
5. Digestion, absorption, and metabolism, p. 296

6. Parietal peritoneum, p. 296
7. Mouth, anus, p. 296
8. Lumen, p. 296
9. Mucosa, p. 296
10. Submucosa, p. 296
11. Peristalsis, p. 296
12. Serosa, p. 296
13. Mesentery, p. 296

Choose the correct answer

14. A, p. 297
15. B, p. 297
16. B, p. 297
17. A, p. 297
18. A, p. 297
19. A, p. 297
20. A, p. 297
21. A, p. 297
22. B, p. 297
23. B, p. 297
24. B, p. 297
25. B, p. 297

Multiple choice

26. E, p. 298
27. C, p. 299
28. E, p. 300
29. D, p. 301
30. B, p. 301
31. C, p. 300
32. D, p. 300
33. D, p. 300
34. D, p. 302
35. B, p. 303
36. C, p. 303
37. A, p. 303
38. A, P. 300
39. B, p. 300
40. C, p. 300

Fill in the blanks

41. Pharynx, p. 303
42. Esophagus, p. 303
43. Stomach, p. 303
44. Cardiac sphincter, p. 304
45. Chyme, p. 304
46. Fundus, p. 305
47. Body, p. 305
48. Pylorus, p. 305

49. Pyloric sphincter, p. 305
50. Small intestine, p. 305

Matching

51. D, p. 305
52. J, p. 305
53. G, p. 303
54. A, p. 303
55. H, p. 305
56. B, p. 304
57. C, p. 305
58. E, p. 305
59. I, p. 305
60. F, p. 305

Multiple choice

61. C, p. 305
62. B, pp. 305 and 307
63. A, p. 307
64. A, p. 308
65. B, p. 307
66. E, p. 307
67. D, p. 305
68. D, p. 307; review Chapter 9 (hormones circulate in blood)
69. B, p. 307
70. C, p. 307

True or false

71. Vitamin K, p. 310
72. No villi are present in the large intestine, p. 310
73. Diarrhea, p. 310
74. Cecum, p. 311
75. Hepatic, p. 311
76. Sigmoid, p. 311
77. T
78. T
79. Parietal, p. 312
80. Mesentery, p. 312

Multiple choice

81. B, p. 313
82. D, p. 313
83. C, p. 313
84. C, p. 313
85. C, p. 313

86. Fill in the blank areas on the chart below.

CHEMICAL DIGESTION

DIGESTIVE JUICES AND ENZYMES	SUBSTANCE DIGESTED (OR HYDROLYZED)	RESULTING PRODUCT
SALIVA		
	1. Starch (polysaccharide)	
GASTRIC JUICE		2. Partially digested proteins
PANCREATIC JUICE		
		3. Starch, peptides and amino acids
	4. Fats emulsified by bile	
INTESTINAL JUICE		
	5. Peptides	
6. Sucrase		
	7. Lactose	
		8. Glucose

APPLYING WHAT YOU KNOW

87. Ulcer
88. Pylorospasm
89. Basal metabolic rate or protein-bound iodine to determine thyroid function

90. WORD FIND

```
. . . . . . . . N . . . . . . . . . U . E
. . . . . . . N X O . . . . . . . . V . C
. . . . . . N . R I I . . . . . . . U . I
. S . P F . O . . U D T . . . . . . L . D
. A . A E . I . . . B N A . . . . . A . N
. E . P C . T D . . . T E C . . . . . . U
D R . I E . S E . . . . R P I . . . . . A
I C . L S . E N . . . . . A P T . . . . J
A N C L . . G T . . . . . F E A S . . P
R A A A . . I I . . . . U . . H . A . E
R P V E N . D N . . . N . . . E . . M R
H . I . . W . . D . D . . . . M . . M I
E . T . . . O U . U . A S O C U M . E S
A . Y . . . O R S . . . . . . L . . S T
. . . . . D . . C . . . . . . S . . E A
. . . N E N O I T P R O S B A I . . N L
. . . N . S T O M A C H . . . F . . T S
. . U . . . . . . . . . . . . Y . . E I
. M . . . . . . . . . . . . . . . . R S
. M E T A B O L I S M . . . . . . . Y .
```

CROSSW

Digestive Organs

1. Parotid gland
2. Pharynx
3. Esophagus
4. Diaphragm
5. Spleen
6. Splenic flexure
7. Stomach
8. Transverse colon
9. Descending colon
10. Ileum
11. Sigmoid colon
12. Rectum
13. Anal canal
14. Vermiform appendix
15. Cecum
16. Ileocecal valve
17. Ascending colon
18. Pancreas
19. Duodenum
20. Common bile duct
21. Hepatic bile duct
22. Hepatic flexure
23. Cystic duct
24. Gallbladder
25. Liver
26. Trachea
27. Larynx
28. Submandibular gland
29. Sublingual gland
30. Tongue

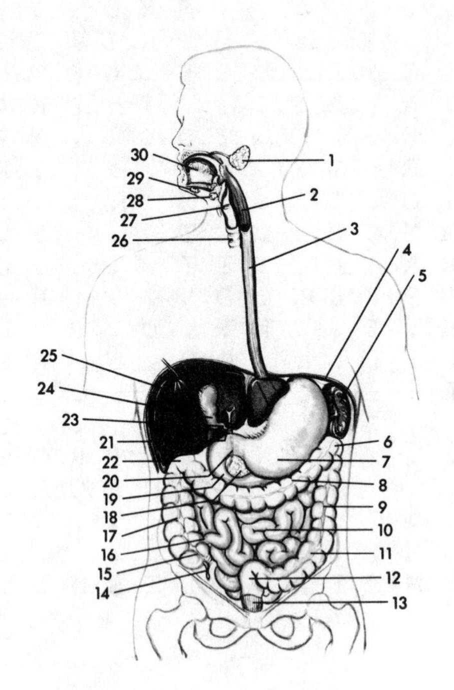

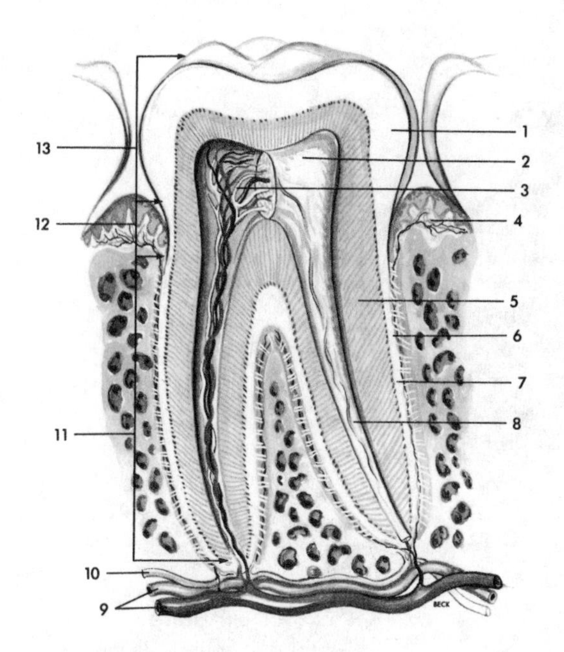

Tooth

1. Enamel
2. Pulp
3. Pulp cavity
4. Gingiva (gum)
5. Dentin
6. Periodontal membrane
7. Cementum
8. Root canal
9. Vein and artery
10. Nerve
11. Root
13. Neck
14. Crown

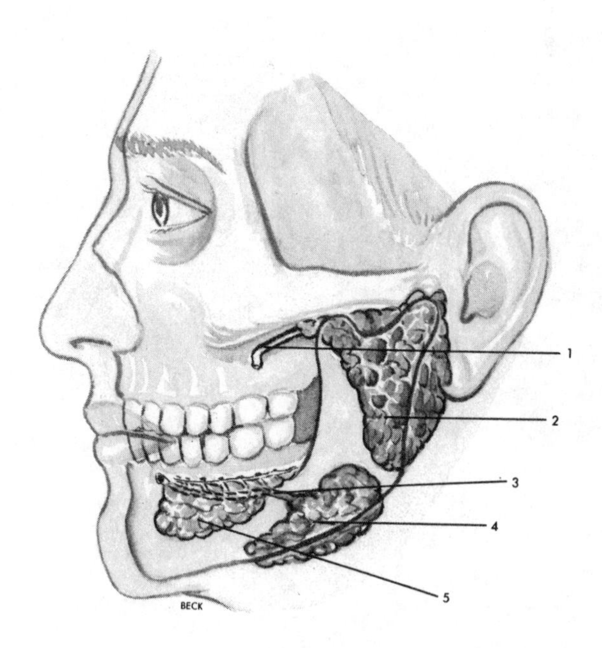

The Salivary Glands

1. Parotid duct
2. Parotid gland
3. Submandibular duct
4. Submandibular gland
5. Sublingual gland

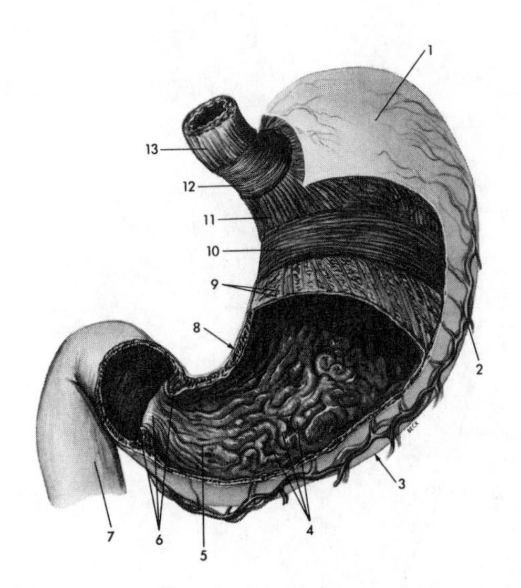

Stomach

1. Fundus
2. Body
3. Greater curvature
4. Rugae
5. Pylorus
6. Pyloric sphincter
7. Duodenum
8. Lesser curvature
9. Oblique muscle layer
10. Circular muscle layer
11. Longitudinal muscle
12. Cardiac sphincter
13. Esophagus

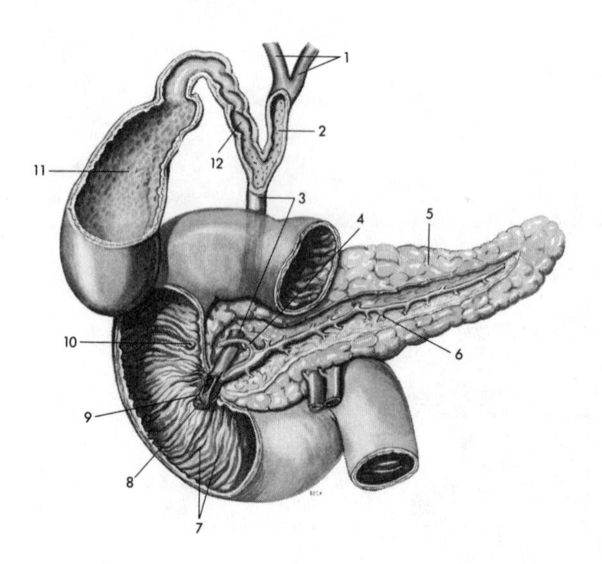

Gallbladder and Bile Ducts

1. Right and left hepatic ducts
2. Common hepatic duct
3. Common bile duct
4. Accessory duct
5. Pancreas
6. Pancreatic duct
7. Duodenum
8. Major duodenal papilla
9. Sphincter muscles
10. Minor duodenal papilla
11. Gallbladder
12. Cystic duct

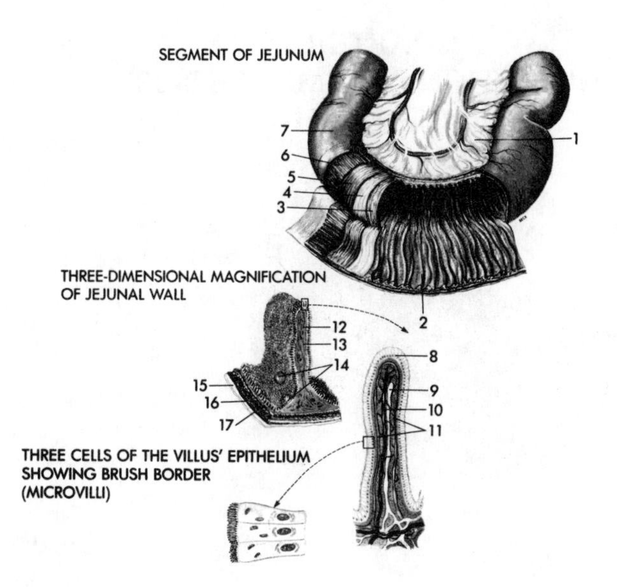

The Small Intestine

1. Mesentery
2. Plica
3. Mucosa
4. Submucosa
5. Circular muscle
6. Longitudinal muscle
7. Serosa
8. Epithelium of villus
9. Lacteal
10. Artery
11. Vein
12. Plica
13. Submucosa
14. Lymph nodules
15. Serosa
16. Circular muscle
17. Longitudinal muscle

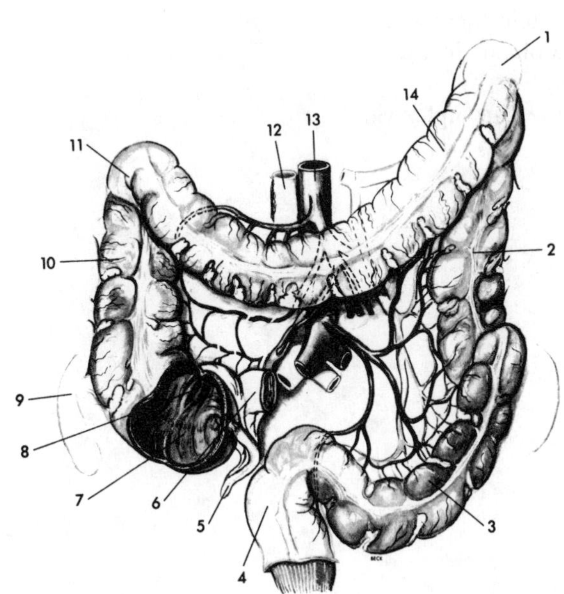

The Large Intestine

1. Splenic flexure
2. Descending colon
3. Sigmoid colon
4. Rectum
5. Vermiform appendix
6. Ileum
7. Cecum
8. Ileocecal valve
9. Ilium
10. Ascending colon
11. Hepatic flexure
12. Inferior vena cava
13. Aorta
14. Transverse colon

CHAPTER 15
NUTRITION AND METABOLISM

Fill in the blanks

1. Bile, p. 320
2. Prothrombin, p. 320
3. Fibrinogen, p. 320
4. Iron, p. 320
5. Hepatic portal vein, p. 320

Matching

6. B, p. 323
7. A, p. 320
8. C, p. 323
9. D, p. 324
10. E, p. 324
11. A, p. 320
12. E, p. 324
13. A, p. 320

Circle the term that does not belong

14. Bile (all others refer to Carbohydrate Metabolism)
15. Amino acids (all others refer to Fat Metabolism)
16. M (all others refer to Vitamins)
17. Pyruvic acid (all others refer to Protein Metabolism)
18. Insulin (all others tend to increase blood glucose)
19. Folic acid (all others are Minerals)
20. Ascorbic acid (all others refer to the B-complex vitamins)

Circle the correct choice

21. C, p. 325
22. A, p. 326
23. C, p. 327
24. B, p. 327
25. B, p. 328
26. A, p. 328
27. C, p. 328
28. D, p. 328
29. A, p. 328

Unscramble the words

30. Liver
31. Catabolism
32. Amino
33. Pyruvic
34. Evaporation

APPLYING WHAT YOU KNOW

35. Weight loss; Anorexia nervosa
36. Iron
    Meat, eggs, vegetables and legumes

37. **WORD FIND**

```
M I N E R A L S . . . . . . . . . . . .
. . . . C . . . . . C O N V E C T I O N
. . . . . A . . I B P . . . R M T . . .
. P . . . . R . . . . . . . . . . . C .
. T . . . . . B . R A D I A T I O N A .
. A . . . . . . O . . . . . . . . . T .
. . . . . . . . . H . . . . . . . . A .
. . . . . . . . . . Y . . . . . . . B .
N . F . . . . . . . . D . . . . . . O .
O . . A . . . . . . L O R E C Y L G L .
I . . . T . . . B M R . . A . . . . I .
T . . . . S . . . R E V I L T . . . S .
C . . . . . . . . . . . . . . E . . M .
U . . . . . . . . . . . . . . . S . . .
D S N I E T O R P . . . . . . . . . . .
N . . . . . . . . A D I P O S E . . . .
O . . . . . . . N O I T A R O P A V E .
C . . . . . . . . . E L I B . . . . . .
. . . . . S I S Y L O C Y L G . . . . .
. . . . . . . . V I T A M I N S . . . .
```

CROSSWORD

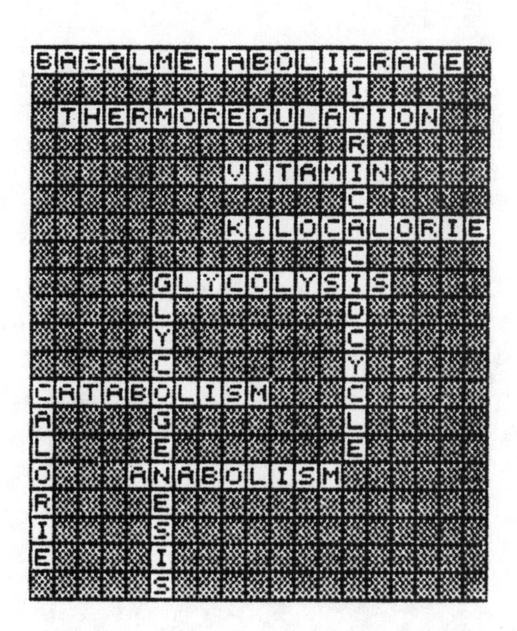

CHAPTER 16
URINARY SYSTEM

Multiple choice

1. E, p. 332
2. C, p. 332
3. E, p. 332
4. C, p. 337
5. E, p. 339
6. C, p. 338
7. B, p. 339
8. E, p. 339
9. C, p. 339
10. C, p. 339
11. B, p. 341 (review Chapter 9)
12. D, p. 341

Choose the correct term

13. G, p. 332
14. I, p. 339
15. H, p. 332
16. B, p. 332
17. K, p. 332
18. F, p. 332
19. J, p. 332
20. D, p. 332
21. L, p. 334
22. C, p. 332
23. M, p. 340
24. A, p. 332

Choose the correct term

25. B, p. 343
26. C, p. 344
27. A, p. 342
28. B, p. 342
29. C, p. 344
30. C, p. 344
31. A, p. 342
32. C, p. 343
33. B, p. 343
34. A, p. 342
35. B, p. 343

Fill in the blanks

36. Renal colic, p. 343
37. Mucous membrane, p. 342
38. Renal calculi, p. 344
39. Ultrasound, p. 344
40. Catheterization, p. 345
41. Cystitis, p. 345
42. Semen, p. 344
43. Urinary meatus, p. 344

Fill in the blanks

44. Micturition, p. 344
45. Urination, p. 344
46. Voiding, p. 344
47. Internal urethral, p. 344
48. Exit, p. 344
49. Urethra, p. 344
50. Voluntary, p. 344
51. Emptying reflex, p. 345
52. Urethra, p. 345
53. Retention, p. 345
54. Suppression, p. 345
55. Automatic bladder, p. 345

APPLYING WHAT YOU KNOW

56. Polyuria
57. Residual urine is often the cause of repeated cystitis

58. **WORD FIND**

```
. . . . . . . . . . . . . . . . . . . .
. . N E P H R O N S I T I T S Y C . Y .
U . . . . . . . A L L U D E M . . . E .
R . . . . P . . . . C O R T E X . . N .
E . . . E . . . . . . H . . . . . . D .
T . . L . . . . . . E . . . . . . . I .
E . V . . . . . . M G . . . . . . . K .
R I . . . . . . O F I L T R A T I O N .
S . . . . . . D . . . P O . . . . R . .
. . X . . . I . . . . Y . M . . . E . .
. . . Y . A A D H . P R . . E . . D . .
. . . . L . . . . A . A C . . R . D . .
. . . Y . A . . P . . M . A . . U A . .
. . S . . . C I . . . I . . L . . L . .
. I . . . . L . . . . D . . . C . B U .
S . . . . L . . . . . S . . . . U . . S
. . . . A . . . . . . . . . . . . L . .
. . . . . . . . . . . . . . . . . . I .
. . . . E C N E N I T N O C N I . . . .
. . . M I C T U R I T I O N . . . . . .
```

CROSSWORD

								L											
C	A	T	H	E	T	E	R	I	Z	A	T	I	O	N					
								T											
								H							C	A	L	Y	X
								O							Y				
	M	I	C	T	U	R	I	T	I	O	N				S				
		N						R							T				
		C				O		I					T	R	I	G	O	N	E
		O				L		P							T				
		N		P		I		T					G		I				
		T		O		G	L	O	M	E	R	U	L	U	S				
		I		L		U		R					Y						
		N		Y		R							C						
		E		U		I							O						
	A	N	U	R	I	A							S						
		C		I									U						
		E		A									R						
													I						
													A						

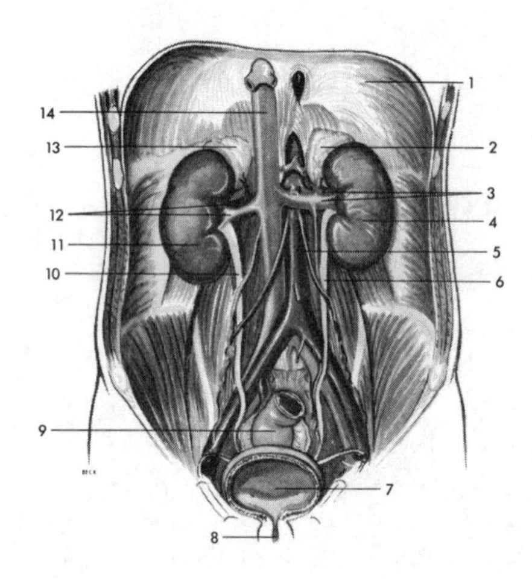

Urinary System

1. Diaphragm
2. Left adrenal gland
3. Left renal artery and vein
4. Left kidney
5. Aorta
6. Left ureter
7. Urinary bladder
8. Urethra
9. Rectum
10. Right ureter
11. Right kidney
12. Right renal artery and vein
13. Right adrenal gland
14. Inferior vena cava

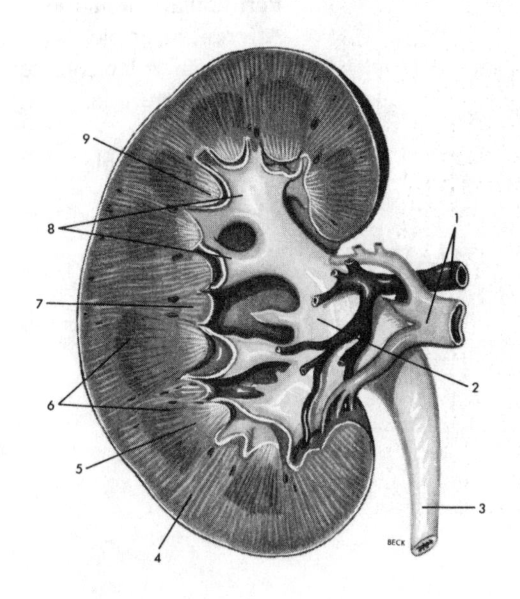

Kidney

1. Renal artery and vein
2. Pelvis
3. Ureter
4. Cortex
5. Pyramid
6. Medulla
7. Renal column
8. Calyx
9. Papilla

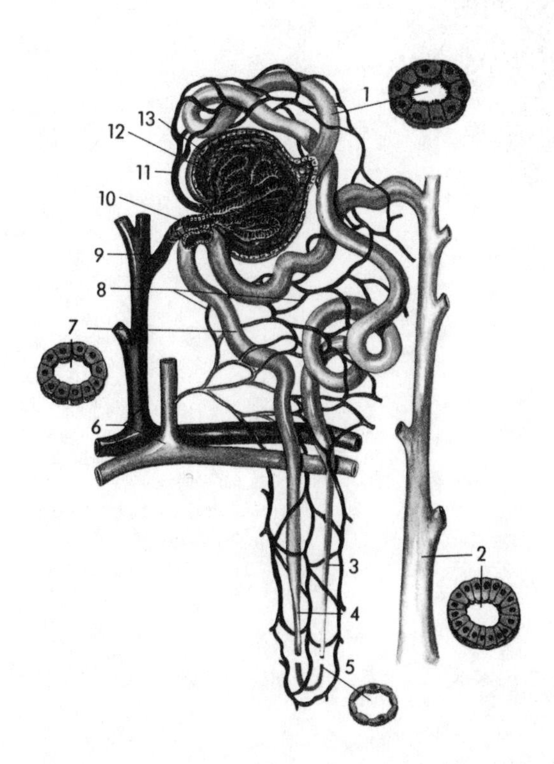

Nephron

1. Proximal convoluted tubule
2. Collecting tubule
3. Descending limb of Henle's loop
4. Ascending limb of Henle's loop
5. Segment of Henle's loop
6. Artery and vein
7. Distal convoluted tubule
8. Peritubular capillaries
9. Afferent arteriole
10. Juxtaglomerular complex
11. Efferent arteriole
12. Glomerulus
13. Bowman's capsule

CHAPTER 17
FLUID AND ELECTROLYTE BALANCE

Circle the correct response

1. Inside, p. 350
2. Extracellular, p. 350
3. Extracellular, p. 350
4. Lower, p. 350
5. More, p. 350
6. Decline, p. 350
7. Less, p. 350
8. Decreases, p. 350
9. 55%, p. 350
10. Fluid balance, p. 350

Multiple choice

11. C, p. 355
12. E, p. 354
13. A, p. 355
14. A, p. 355
15. C, p. 353
16. E, p. 353
17. D, p. 353
18. D, p. 353
19. C, p. 357
20. B, p. 357
21. D, p. 355
22. E, p. 358
23. B, p. 357
24. B, p. 359
25. B, p. 359
26. E, p. 357

True or false

27. Catabolism, p. 350
28. T
29. T
30. Nonelectrolyte, p. 354

31. T
32. Hypervolemia, p. 356
33. Tubular function, p. 355
34. 2400 ml, p. 353
35. T
36. 100 mEg, p. 357

Fill in the blanks

37. Dehydration, p. 359
38. Decreases, p. 359
39. Decrease, p. 359
40. Overhydration, p. 359
41. Intravenous fluids, p. 359
42. Heart, p. 359

APPLYING WHAT YOU KNOW

43. Ms. Titus could not accurately measure water intake created by foods or catabolism, nor could she measure output created by lungs, skin, or the intestines.

44. A careful record of fluid intake and output should be maintained and the patient should be monitored for signs and symptoms of electrolyte and water imbalance.

CROSSWORD

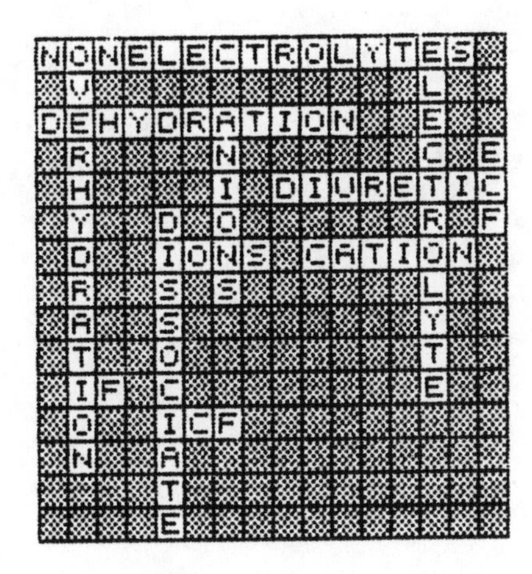

CHAPTER 18
ACID-BASE BALANCE

Choose the correct term

1. B, p. 364
2. A, p. 364
3. A, p. 364
4. B, p. 364
5. B, p. 364
6. B, p. 364
7. B, p. 365
8. A, p. 365
9. B, p. 365
10. B, p. 365

Multiple choice

11. E, p. 364
12. E, p. 364
13. A, p. 365
14. E, p. 368
15. C, p. 369
16. D, p. 372
17. C, p. 369
18. B, p. 369
19. D, p. 369
20. E, p. 369
21. E, p. 370

True or false

22. Buffer instead of heart, p. 364
23. Buffer pairs, p. 365
24. T
25. T
26. Alkalosis, p. 369
27. Kidneys, p. 369
28. T
29. Kidneys, p. 372
30. Lungs, p. 372

Matching

31. E, p. 372
32. G, p. 372
33. F, p. 372
34. A, p. 372
35. I, p. 373

36. B, p. 372
37. H, p. 372
38. C, p. 372
39. D, p. 372
40. J, p. 372

APPLYING WHAT YOU KNOW

41. Normal saline contains chloride ions which replace bicarbonate ions and thus relieve the bicarbonate excess which occurs during severe vomiting.

42. Most citrus fruits, although acid tasting, are fully oxidized during metabolism and have little effect on acid-base balance. Cranberry juice is one of the few exceptions.

43. Milk of magnesia. It is base. Milk is slightly acidic. (see chart, p. 365)

44. **WORD FIND**

```
. . . . . . . . . . . . . . . . . . N .
. . D I U R E T I C . . . . . . . . O .
D . . . . E C N A L A B D I U L F . I .
E . . S . . A L D O S T E R O N E . T .
H E . . N . . . . . . . . . . . . . A .
Y L . . . O . . . . . . . . . . . H R .
D E . . . . I . . S . . . . . . O . D .
R C . . . . . T O . . . K . . M . . Y .
A T . . . . . D A . . . . I E . . . H .
T R . . . . I . . C . I . O D . . . R .
I O . . . U A . . O . N S . . N . . E .
O L . . M N . . U . . T . . . . E . V .
N Y . . I . . T . . A A . . . . D Y O .
. T . O . . P . . S . K T W . . E . S .
. E N . . U . . I . . E S . A . M . . .
. S . . T . . S . . . . R . . T A . . .
. . . . . . . . H . . . I . . . E . . .
. . . . . . . D . . . . H . . . . R . .
. . . . . . A . . . . . T . . . . . . .
. N O N E L E C T R O L Y T E S . . . .
```

CROSSWORD

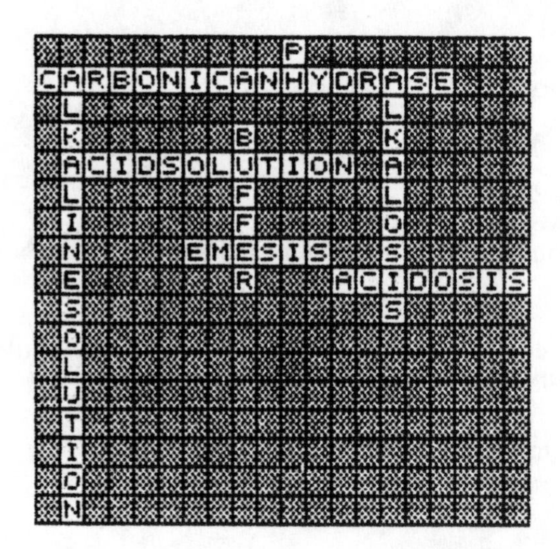

CHAPTER 19
THE REPRODUCTIVE SYSTEMS

Matching

Group A

1. D, p. 378
2. C, p. 378
3. E, p. 378
4. B, p. 379
5. A, p. 378

Group B

6. C, p. 378
7. A, p. 379
8. D, p. 378
9. B, p. 379
10. E, p. 379

Multiple choice

11. B, p. 379
12. C, p. 381
13. A, p. 381
14. D, p. 381
15. E, p. 383
16. D, p. 383
17. C, p. 384
18. C, p. 381
19. A, p. 383
20. B, p. 381

Fill in the blanks

21. Testes, p. 379
22. Spermatozoa or sperm, p. 381
23. Ovum, p. 383
24. Testosterone, p. 383
25. Interstitial cells, p. 383
26. Masculinizing, p. 384
27. Anabolic, p. 384

Choose the correct term

28. B, p. 384
29. H, p. 380
30. G, p. 385
31. A, p. 384
32. F, p. 384
33. C, p. 384
34. I, p. 384
35. E, p. 384
36. D, p. 386
37. J, p. 385

Matching

38. D, p. 386
39. C, p. 386
40. B, p. 386
41. A, p. 386
42. E, p. 386

Choose the correct structure

43. A, p. 392
44. B, p. 389
45. A, p. 392
46. B, p. 389
47. A, p. 392
48. A, p. 392
49. A, p. 392
50. B, p. 389

Fill in the blanks

51. Gonads, p. 386
52. Oogenesis, p. 386
53. Meiosis, p. 388
54. One half or 23, p. 388
55. Fertilization, p. 388
56. 46, p. 388

57. Estrogen, p. 388
58. Progesterone, p. 388
59. Secondary sexual characteristics, p. 388
60. Menstrual cycle, p. 388
61. Puberty, p. 388

Choose the correct structure

62. A, p. 388
63. B, p. 390
64. C, p. 390
65. B, p. 390
66. A, p. 388
67. B, p. 389
68. A, p. 389
69. A, p. 388
70. C, p. 390
71. B, p. 390

Matching

Group A

72. D, p. 390
73. E, p. 390
74. B, p. 391
75. C, p. 391
76. A, p. 391

Group B

77. E, p. 391
78. A, p. 391
79. D, p. 391
80. B, p. 391
81. C, p. 392

True or false

82. Menarche, p. 392
83. One, p. 395
84. 14, p. 395
85. Menstrual period, p. 395
86. T
87. Anterior, p. 395

Choose the correct hormone

88. B, p. 395
89. A, p. 395
90. B, p. 395
91. B, p. 395
92. A, p. 395

APPLYING WHAT YOU KNOW

93. Impotent - The testes are not only essential organs of reproduction, but are also responsible for the "masculinizing" hormone. Without this hormone, Mr. Belinki will have no desire to reproduce.

94. Sterile - The sperm count may be too low to reproduce but the remaining testicle will produce enough masculinizing hormone to prevent impotency.

95. The uterine tubes are not attached to the ovaries and infections can exit at this area and enter the abdominal cavity.

96. Yes. Yes. Without the hormones from the ovaries to initiate the menstrual cycle, Mrs. Harlan will no longer have a menstrual cycle and can be considered to be in menopause. (cessation of menstrual cycle)

97. No. Ms. Comstock still will have her ovaries which are the source of her hormones. She will not experience menopause due to this procedure.

98. **WORD FIND**

```
. . . . M . . . . . . . . . . . . . . .
. . . . U . . . . . . . . . . . . . . .
A . . . T . O V I D U C T S . . . . . .
C . . . O . . . . . . . C O W P E R S M
R . . . R . . . . . . M . . . . . . . S
O . . . C . . . . . . . E . . . . . . I
S S . . S . . . . . . . . I . . . . . D
O N S E M I N I F E R O U S O . . . . I
M E . . E N D O M E T R I U M S . . S H
E R . . . . . . . . . . . . . . I . I C
. E . . . . . . S P E R M A T I D S M R
. F . S . E S T R O G E N . . S . . Y O
. E . E . . . . . . . . . . . P . . D T
. D . I . . . . . . . . . . . E . V I P
. S P R O S T A T E C T O M Y R . A D Y
. A . A . . P . . . . . . . . M . G I R
. V . V . . E . . . . . . . . . . I P C
. . . O . . N . . . . . . . . . . N E .
. . . . . . I . . . . . . . . . . A . .
. . . . . . S P R E G N A N C Y . . . .
```

CROSSWORD

			V														
			U														
			L														
			V														
G	R	A	A	F	I	A	N	F	O	L	L	I	C	L	E		
		R											I				
		E			T	E	S	T	O	S	T	E	R	O	N	E	
		O											C				
	C	L	I	T	O	R	I	S					U				
		A									G	A	M	E	T	E	S
											O		C				
											N		I				
							P				A		S	E	M	E	N
							R				D		I				
						M	E	N	S	E	S		O				
							P						N				
							U										
							C										
							E										

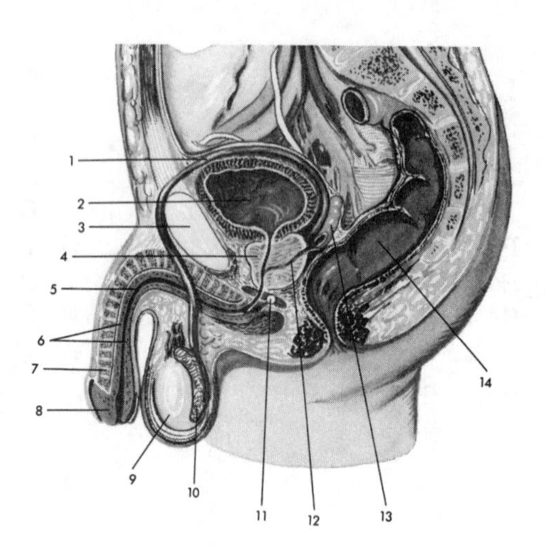

Male Reproductive Organs

1. Ductus deferens
2. Bladder
3. Symphysis pubis
4. Prostate gland
5. Urethra
6. Corpus spongiosum urethra
7. Corpus cavernosum
8. Glans penis
9. Testis
10. Epididymis
11. Bulbourethral gland
12. Ejaculatory duct
13. Seminal vesicle
14. Rectum

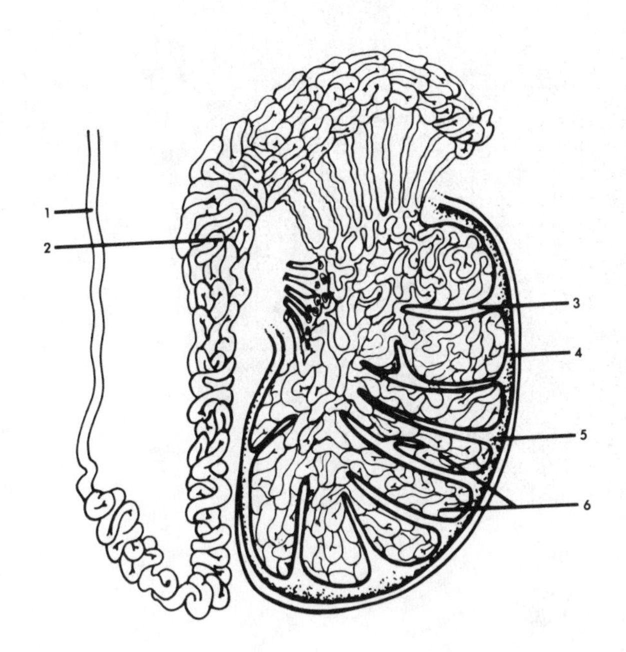

Tubules of Testis and Epididymis

1. Ductus (vas) deferens
2. Body of epididymis
3. Septum
4. Lobule
5. Tunica albuginea
6. Seminiferous tubules

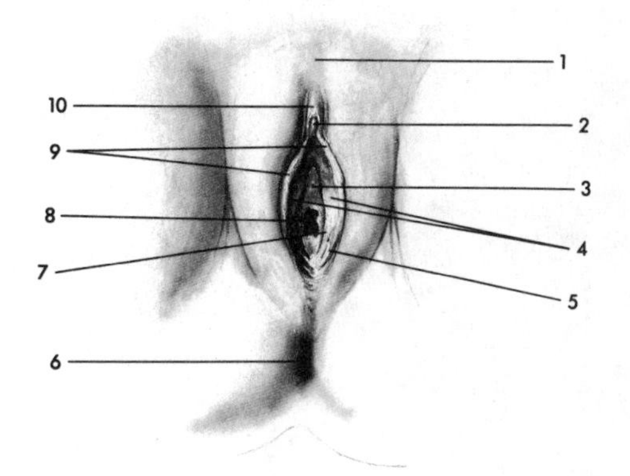

Vulva

1. Mons pubis
2. Clitoris
3. Orifice of urethra
4. Labia majora
5. Opening of greater vestibular gland
6. Anus
7. Vestibule
8. Orifice of vagina
9. Labia minora
10. Prepuce

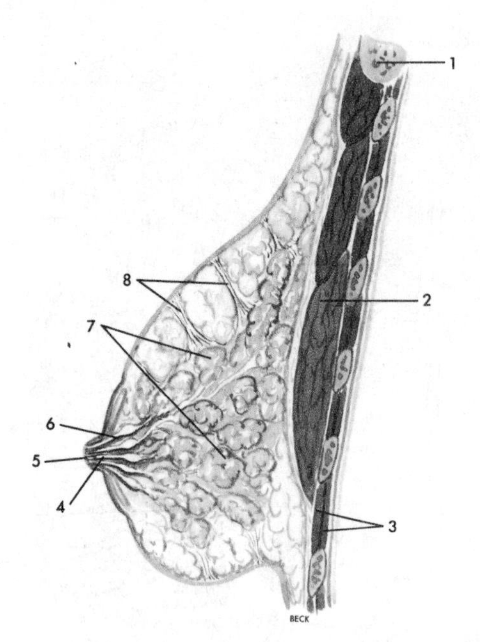

Breast

1. Clavicle
2. Pectoralis major muscle
3. Intercostal muscles
4. Duct
5. Nipple
6. Lactiferous duct
7. Alveoli
8. Suspensory ligaments of Cooper

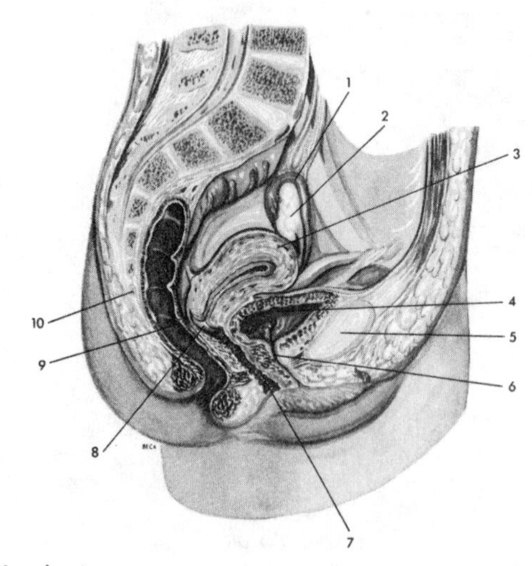

Female Pelvis

1. Fallopian tube (Uterine)
2. Ovary
3. Uterus
4. Urinary bladder
5. Symphysis pubis
6. Urethra
7. Vagina
8. Cervix
9. Rectum
10. Coccyx

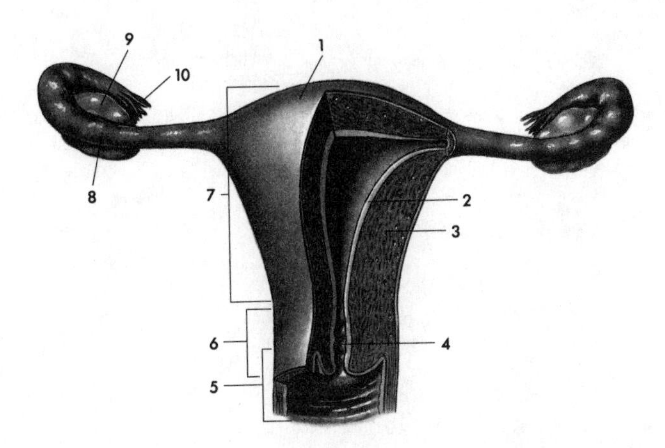

Uterus and Adjacent Structures

1. Fundus
2. Endometrium
3. Myometrium
4. Cervical canal
5. Vagina
6. Cervix
7. Body
8. Uterine (Fallopian) tube
9. Ovary
10. Fimbriae

CHAPTER 20
GROWTH AND DEVELOPMENT

Fill in the blanks

1. Conception, p. 402
2. Birth, p. 402
3. Embryology, p. 402
4. Oviduct, p. 402
5. Zygote, p. 402
6. Morula, p. 402
7. Blastocyst, p. 402
8. Amniotic cavity, p. 406
9. Chorion, p. 406
10. Placenta, p. 406

Choose the correct term

11. G, p. 407
12. F, p. 410
13. C, p. 412
14. B, p. 409
15. A, p. 407
16. H, p. 410
17. E, p. 410
18. D, p. 410
19. I, p. 409
20. J, p. 412

Multiple choice

21. E, p. 413
22. E, p. 413
23. E, p. 414
24. A, p. 414
25. B, p. 414
26. C, p. 414
27. B, p. 414
28. D, p. 414
29. E, p. 414
30. D, p. 414
31. A, p. 415
32. C, p. 415
33. C, p. 415
34. C, p. 415
35. E, p. 415

Matching

36. F, p. 412
37. A, p. 413
38. C, p. 415
39. H, p. 414
40. D, p. 414
41. B, p. 413
42. E, p. 415
43. G, p. 413
44. I, p. 415

Fill in the blanks

45. Lipping, p. 415
46. Osteoarthritis, p. 417
47. Nephron, p. 417
48. Barrel chest, p. 417
49. Atherosclerosis, p. 417
50. Arteriosclerosis, p. 417
51. Hypertension, p. 417
52. Presbyopia, p. 417
53. Cataract, p. 418
54. Glaucoma, p. 418

Unscramble the words

55. Infancy
56. Postnatal
57. Organogenesis
58. Zygote
59. Childhood
60. Fertilization

APPLYING WHAT YOU KNOW

61. Normal
62. Only about 40% of the taste buds present at ago 30 remain at age 75
63. A significant loss of hair cells in the Organ of Corti causes a serious decline in ability to hear certain frequencies

CROSSWORD

												N							
		A	R	T	E	R	I	O	S	C	L	E	R	O	S	I	S		
		T										O							
	C	H	O	R	I	O	N			S	E	N	E	S	C	E	N	C	E
		E										A				M			
M	O	R	U	L	A		G					T				B			
		O					L		H			E				R			
		S			C		A		I							Y			
		C			A		U		S				Z	Y	G	O	T	E	
	B	L	A	S	T	O	C	Y	T	E						L			
		E			A		O		O			P				O			
		R			R		M		G			R				G			
		O			A		A		E			E				Y			
		S			C				N			S							
		I			T				E			B							
		S							S			Y							
			P	A	R	T	U	R	I	T	I	O	N						
									S			P							
												I							
												A							

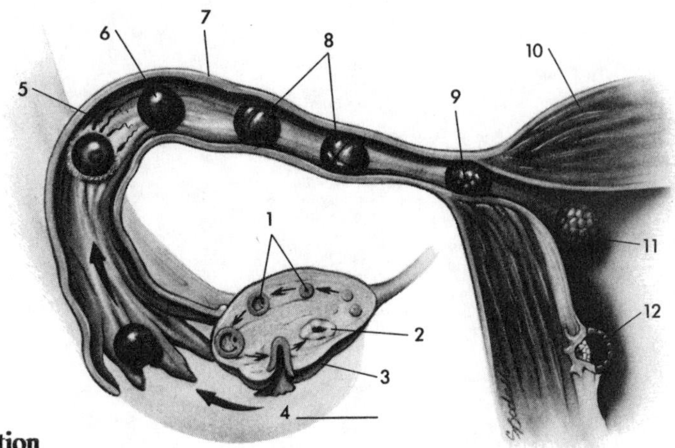

Fertilization and Implantation

1. Developing follicles
2. Corpus luteum
3. Ovary
4. Ovulation
5. Spermatozoa (Fertilization)
6. First mitosis
7. Uterine (Fallopian) tube
8. Divided zygote
9. Morula
10. Uterus
11. Blastocyst
12. Implantation

NOTES

NOTES

NOTES